# THE WASTE-WISE GARDENER

## TIPS AND TECHNIQUES TO SAVE TIME, MONEY, AND NATURAL RESOURCES WHILE CREATING THE GARDEN OF YOUR DREAMS

(Low-Cost, Easy-Care Sustainable Gardening)

***Jean B. MacLeod***

MacLeod How-To Books
The Waste-Wise Gardener: Tips and Techniques to Save Time, Money, and Natural Resources While Creating the Garden of Your Dreams; (Low-Cost, Easy-Care Sustainable Gardening)



All ideas, suggestions, and guidelines mentioned in this book are written for informational purposes only and should not be considered as recommendations or endorsements for any particular product. The content included is not intended to be, nor does it constitute, the giving of professional advice. All recommendations are made without guarantees on the part of the author. The author shall not be held liable for damages incurred through the use of the information provided herein.



ISBN: 0997446420
ISBN-13: 9780997446425
Library of Congress Control Number: 2017912135
Jean B. MacLeod, Torrance, CA

*In loving memory of my grandmother*
*Annie Elizabeth Blake*
*October 31, 1872–March 26, 1955*

# Acknowledgments

To gardeners everywhere, who make the world a better place.
And to you, dear reader, thank you.

# Contents

# Introduction

I CAME TO GARDENING WHEN my husband and I bought our first home in the early 1960s. Located in an older part of town, the house was a split-level built in the 1920s, with an adjacent terraced garden containing exquisite stonework but not much in the way of vegetation. I was a stay-at-home mom at the time, and the garden presented a wonderful opportunity to dive right in and try my chops at gardening. The property was a perfect palette just waiting for a gardener's touch. I went for gardening with a passion.

Leaving no stone unturned in nurturing my new love, I pored over gardening books and catalogs, soaked myself in gardening lore, and concentrated on building up the clay-like soil. It was an exhilarating experience, even adding to the compost heap: manure and straw from riding stables, grass clippings and leaves from neighbors, and washed-up seaweed from the beach. Wherever there was material I could use for the garden, I was Johnny-on-the-spot. Nothing was overlooked.

My garden blossomed, and I reveled in every minute of my horticultural quest—weeding, digging, mulching. I planted every inch of ground. Passersby commented on the flower display, my family reveled in fresh veggies and fruit, and even the cat luxuriated in its own catnip plant. I was Mrs. Green Thumb personified. Or so I thought. All this was not without mistakes, however: a weeping willow was doomed from the start, a plant's biodegradable container just didn't biodegrade, and there were plant youngsters I killed with kindness. I should have known better.

Nothing lasts forever. The euphoria fizzled out somewhat when I took a position outside the home. I was in the same predicament as many other time-strapped gardeners who probably feel that the garden represents more labor than love, that its maintenance and problems outweigh its pleasures. When the lawn is languishing, the water bill burgeoning, and the pests proliferating, a garden does seem more bother than bliss. What happened to the joy of gardening? Did it fly out the window to be replaced with a never-ending to-do list: feed the flowers, prune the perennials, and whittle down the weeds? Plus, what about saving seeds, conserving water, recycling and reusing, and being an ethical gardener in our changing environment?

Times have changed in the gardening world since the 1960s, and gardeners have kept pace with the times. We read. We stay informed. But some things never change, such as the desire to enjoy our gardens, not just labor in them, which is how the idea for this book originated. We *can* have our cake and eat it too. We can have gardens that relieve stress rather than create it, flourish without constant attention, and are sources of personal joy and fulfilment.

Arranged alphabetically, *The Waste-Wise Gardener* provides information on garden plants, garden pests, and gardening conundrums, along with practical ways to save time, money, and natural resources. Whether you're a beginner or a seasoned gardener, you'll find the book a handy resource for day-to-day gardening challenges, along with plenty of ways to minimize work and maximize benefits. Whether you wish to grow flowers or vegetables, nurture a potted plant, or simply have more time to stand and stare, you'll find both inspiration and information aplenty.

By using planet-friendly waste materials, selecting the proper plants, cooperating with nature, and dealing with gardening problems in a practical fashion, it's a short step to creating your ideal garden: an inviting space that wastes less and contributes more; a habitat haven for pollinators; an eco-friendly environmental asset; and a low-cost, low-maintenance little spot of heaven—right outside your door.

# A

***ANIMALS, to deter;*** see also *CORN, to protect from raccoons; CATS, to deter; RABBITS, to repel.*

- Keep dogs, raccoons, and opossums from getting into garbage cans by spraying a diluted ammonia solution on the cans or pouring a little undiluted ammonia into the can itself. Another option is to sprinkle a few tablespoons of Epsom salt around the cans.
- Prevent animals from tipping over garbage cans by tying the cans together with bungee cords. Also use the cords to attach the cans to a fence and to secure the lids to the cans.

***ANNUALS, EASIEST TO GROW FROM SEED***

- The easiest annuals to grow from seeds are sunflower, blue honeywort, breadseed poppy, California poppy, calliopsis, catchfly, cleome, coleus, cornflower, cosmos, Indian blanket, larkspur, marigold, peony poppy, red poppy, sunflower, and zinnia.

***APHIDS, to control;*** see also *PESTS, PLANT, to control*

- Wash them off with a strong spray of water from the garden hose, hitting both sides of the leaves. Repeat every couple of days.
- Spray the plants with insecticidal soap or pepper spray. Or set out yellow sticky traps (see *INSECTICIDAL SOAP; PEPPER SPRAY; STICKY INSECT TRAPS, to make*).
- Lay aluminum foil under the affected plant. Aluminum foil also repels flea beetles.

***ASPARAGUS, hints on***

- Plant asparagus with tomatoes, parsley, and basil, which benefit the growth of asparagus and control the asparagus beetle.
- Start checking for asparagus beetles after asparagus plants emerge. If you do find any beetles, hold a can of soapy water under the plant, and touch any beetles you find. The beetles will drop into the can of water.
- Obtain white asparagus by covering the spears to prevent them from receiving sunlight and producing chlorophyll, which is what makes them green. Cover them with soil when they start poking through the ground. Keep mounding the soil on top of them as they grow. Another alternative is to cover the emerging spears with inverted clay pots. Cover the drainage holes with rocks to keep out light, and check the spears every few days.

***AZALEAS, hints on***

- Purchase azaleas while they're in bloom so you know what they look like, but don't feed them until they have finished blooming.
- Choose white if you want the longest-lasting azaleas.
- Pinch off the growing tips of evergreen azaleas to promote bushy plants and more blooms.
- Sprinkle with a fine mist of water throughout the growing season. Azaleas thrive on overhead watering. Make sure, however, that the roots are also watered.
- Use a whisk broom to dislodge the dead blossoms when the plants have finished blooming. This will also prevent the plant from setting seed.
- Try moving azaleas to a shadier area if their leaves start acquiring brown spots. The plants may be getting too much sun.
- Take three- to five-inch cuttings in the spring. Fill a pot or empty coffee can with equal amounts of moistened cocopeat and sand. Insert the cuttings, and then cover them with an inverted drinking glass or jar or a

plastic bag. Prop up the bag with straws or plant markers inserted into the potting mix, and hold them in place with a rubber band. Set the potting container in a warm, well-lit place but out of direct sunlight. Check for roots after two weeks and then again after another week or so.

# B

**_BASIL, hints on_**

- Plant with tomatoes, sweet peppers, and eggplants, all of which are beneficial companions.
- Keep tops pinched back to avoid spindly growth and to prevent the plant from setting seed. Use what you pinch off for cooking. When the plant does set seed, leave it to self-sow, or cut it about six inches from the ground.
- Cut the stems regularly. The more you cut the plant, the more it will grow.
- Propagate new plants from vigorous nonflowering stems. Strip the leaves from the lower halves of the stems, insert them in water, and change the water daily.

**_BEANS, hints on_**

- Avoid presoaking or presprouting the seeds; it could cause them to rot. Sow them directly in the ground, water the ground thoroughly, and do not water again until seedlings appear.
- Plant bush beans near potatoes. The potatoes will repel the Mexican bean beetle, and the beans will repel the Colorado potato beetle. Planting an early-bearing bean variety will also avoid the Mexican bean beetles.
- Avoid planting beans close to members of the onion family (chive, garlic, leek, onion, or shallot). They are not beneficial companions.
- Grow pole beans if you're short on space. Pole beans produce three times the yield of bush beans in the same amount of space. Make sure they don't overshadow other plants.
- Train pole beans on a tepee to provide greater air circulation and faster picking. Tie three or four poles or stakes together at the top, or cross

them in the middle and tie them. To give the vines extra support, wrap string around the tepee.

- Plant pole beans with corn after the corn is at least six inches tall. When the beans are ready for support, the corn stalks are tall enough to provide it. Beans also put nitrogen (which corn depletes) back into the soil.
- Repel flea beetles and cucumber beetles by interplanting catnip, tansy, radishes, and nasturtiums.
- Harvest beans almost daily to keep the plants producing. Pinch off bush beans, and use scissors to harvest pole or runner beans. Don't pick beans early in the morning, when the dew is on the vines, or touch the plants when they are wet. This can cause rust and spread disease.

***BEETS, hints on***

- Soak the seeds for twelve hours when planting during hot, dry weather.
- Sow the seeds directly in the ground; they don't transplant well because of the long taproots.
- Place sand, dry coffee grounds, or wood ashes in the bottom of the row or planting hole as a safeguard against root maggots.
- Plant every two weeks for a continuous harvest, and use the tiny beets you thin out to cook or put in salads.
- Deter chipmunks from eating the beets by interplanting onions in the same row. Alternate one beet and one onion. Another alternative is to plant the onions around the beets.
- Plant beets with onions or kohlrabi, which are friends, but not with pole beans, which are not beneficial companions.
- Grow golden beets or white beets if you enjoy beets in salad but don't like the way the red varieties discolor the greens.
- Harvest one-third of the plant's mature leaves throughout the season. This won't hurt the root. Cut the greens, don't pull at them.
- Harvest the roots when they are one and a half to two inches in diameter. This is when they are most tender.

***BENEFICIAL INSECTS;*** see *INSECTS, BENEFICIAL, hints on*

***BIRD FEEDER, HUMMINGBIRD, hints on***

- Choose a bird feeder that has red parts, or hang a red ribbon on the feeder. Hummingbirds are attracted to the color red.
- Clean black mold off the feeder with white vinegar. Avoid using soap or detergent, which might leave a harmful residue.
- Keep ants off the feeder by spreading a little olive oil on the tip that dispenses the nectar. It won't hurt the hummingbirds.
- Make a nectar by mixing four parts water to one part sugar. Boil the water first, and then add the sugar, and let the solution cool before pouring it into the feeder. Be sure to clean the dispenser in very hot water and refill it with fresh syrup every three to four days. Don't use honey to make the syrup; it can cause a fungal growth on hummingbirds' beaks.

***BIRDBATHS, hints on***

- Have the water in the bath be no deeper than three inches in the center, and set the bath at least three feet off the ground.
- Place the bath near shrubs or overhanging branches so the birds have an escape route from cats or other predators.
- Put a few colored marbles in the water to attract the birds.
- Hose out the bath daily, and scrub it with hot soapy water at least once a week. Or use a 10 percent liquid chlorine bleach or distilled white vinegar solution to help kill algae and bacteria. Rinse thoroughly, and let dry before refilling.
- Remove algae on an old birdbath by placing bleach-soaked paper towels on the birdbath and letting them sit for thirty minutes to one hour; remove the paper towels, and then rinse the bath thoroughly, and let it dry.

***BIRDS, to keep away from fruit trees***

- Cut used Mylar balloons into ribbons about one inch wide, and tie them to the tree branches. (Don't put the deterrents in place until the fruit begins to ripen.)

- Loop thread over and around the tree branches to distract the birds.
- Cover small trees with plastic netting. Fasten the protection with clothespins, or tie the ends together below the branches.
- Drape a large tree with several plastic nets sewn together. Use a garden rake to lift it over the tree, and then gather the netting at the base, or anchor it to the ground with stakes.

***BIRDS, to keep away from newly seeded areas***

- Cover the areas with newspapers, burlap, old flannel sheets, or any fabric that will absorb moisture, and keep it moist until the seeds have germinated. Anchor the covering with rocks or a board if necessary. Remove the covering as soon as some of the seeds start to germinate.
- Cover the seeds with grass clippings or a thin layer of straw.
- Crisscross string, heavy thread, wire, or fishing line over the planted area. Use row markers, wooden stakes, chopsticks, or pieces of branches to connect the covering.
- Cover the seeds with chicken wire. Crease the wire down the center vertically to make a shallow tent and place it over the row or seedbed.
- Suspend fine nylon mesh, plastic netting, or old sheer curtains over the areas.

***BIRDS, to keep from eating corn;*** see *CORN, to protect from birds*

***BIRDS, to keep from eating grapes***

- Cover the vines with plastic netting.
- Cover each bunch with a doubled-up mesh bag or a paper bag before the grapes begin to ripen. Cut a hole in the bottom of the paper bag so rainwater does not collect inside.

***BIRDS, to keep from eating ripening melons***

- Place mesh laundry baskets or milk crates over the fruits.
- Place nylon or plastic netting over the melon plants; anchor the netting with stakes or weigh the edges of the netting down with soil.

***BIRDS, to keep from eating sprouts (beans, peas, corn)***

- Prop a berry basket (the kind that strawberries and cherry tomatoes come in) over each planted seed. Remove the baskets after the sprouts reach the tops of the baskets.
- Crisscross string, heavy thread, or wire two inches above the planted area.

***BIRDS, to keep from eating strawberries***

- Suspend plastic netting or cheesecloth over the strawberries. (Save mesh bags that onions or fruit are sold in. Open them up, and string them together.)
- Stick shiny metallic pinwheels in the beds just when the fruit is ripening. Birds don't like the sound or motion pinwheels make. (Don't depend on this if there is no air movement.)
- Paint strawberry-size rocks red, then sprinkle them around the bed before the strawberries ripen.

***BIRDS, to keep out of the garden***

- Stretch a length of cassette tape tightly between stakes placed at either end of a row of plants. The humming noise created by the tape vibrating in the wind can scare off birds. (Look for old cassettes at yard sales or used-record stores.)
- Tie strips of black, red, or orange cloth to posts in the garden. The strips will flap around in the wind and act as a deterrent.

***BIRDS, to welcome to the garden in the winter***

- Fill a paper cup half full of birdseed, and then add any fat left in a frying pan or skimmed from meat dishes. When the cup is full, put it in the refrigerator until the mixture hardens. Peel off the paper cup, and put out the contents for the meat-eating birds. It's a great winter treat for them.
- Save the seeds from squash, sunflowers, and pumpkins, and put them out.

- Add finely crushed eggshells to birdseed, or sprinkle them on bare ground or a rock (wash and bake or microwave the shells to sterilize them first).
- Cover a pinecone with bacon grease, chicken fat, or meat fat, and then roll it in breadcrumbs, rolled oats, or birdseed.
- Use sunflower seeds in the bird feeder, especially black-oil sunflower seeds. Most seed-eating birds favor them above other seeds. Judge the quality of birdseed by the ingredients on the label. Sunflower seeds should always be listed first, followed by other good seeds such as safflower and white millet. Avoid the cheap mixes that contain a lot of red milo, which birds reject.
- Use an old watering can as a handy container for refilling the bird feeder with seed.

***BROCCOLI, hints on***

- Prevent insects from laying their eggs on the broccoli by covering young plants with a polyester row cover or fine cheesecloth. (See *CABBAGE PESTS, to control.* Broccoli is a member of the cabbage family and susceptible to the same pests and diseases.)
- Pick broccoli before the individual florets begin to open so the plants will produce more over a longer period, and pick in early morning for best flavor.
- Harvest the main head, and then keep picking the small clusters from the side branches. The broccoli will keep producing in leaf axis and over the lower stalk. If you don't want too many small side shoots, cut two to four inches of stem when you cut the head.

***BRUSSELS SPROUTS, hints on***

- Maximize space by planting early-maturing vegetables between the sprouts.
- Shield the sprouts from egg-laying insects by covering them with sections of nylon net, fine cheesecloth, or nylon pantyhose. Put the protection in place when the sprouts begin to form.

- Snap off the sprouts from the bottom of the stalk first and leave the sprouts on the upper stem to mature. Start picking the sprouts when they are about the size of marbles, and harvest early in the day, when they will snap off easiest.
- Remove the leaves that grow below each sprout as it matures. This will give the sprouts room to develop.
- Pinch off the plant's top to force sprouts to mature faster, but always leave the top leaves; the plant needs them to supply nourishment.
- Pull up the whole plant (roots and all) when the season is over. Hang it right side up in a cool place, and it will last several weeks.
- See also *CABBAGE PESTS, to control.* Brussels sprouts belong to the same tribe as cabbage and share the same pests and diseases.

***BUCKETS, hints on***

- Buy empty two- or five-gallon plastic buckets from restaurants, bakeries, doughnut shops, or delicatessens. Sometimes the buckets are free for the asking. The five-gallon size makes a fine container for growing cherry tomatoes or smaller vegetables, for making manure or compost tea, or for storing birdseed, fertilizer, and other garden supplies. The two-gallon size works well for mixing up batches of liquid fertilizer. (Use a marker or red fingernail polish to plainly indicate quart and gallon levels inside the bucket.)
- Use buckets with handles to carry small tools and supplies around the garden and to hold weeds, dead wood, and spent flowers while you make the rounds.
- Thoroughly wash, rinse, and dry all buckets and pails before using.

***BULBS, hints on***

- Store bulbs in mesh bags that onions or fruit come in. Hang them high in a cool, dry place. Alternatively, put the dry bulbs in plastic or paper bags (or a box) with peat moss, wood shavings, vermiculite, or shredded newspaper.

- Avoid storing bulbs in the refrigerator if there are apples or other fruit inside. The ethylene gas given off by fruit is harmful to bulbs.
- Remember that daffodils face the sun when selecting a planting site for them.
- Plant bulbs slightly deeper than the recommended guidelines in hot climates or sandy soil, and slightly shallower in heavy soil.
- Save time and effort by planting several bulbs in one large hole. Blooming bulbs also look less formal when planted this way.
- Protect the bulbs from rot by putting bulb food or all-purpose fertilizer in the bottom of each planting hole, and covering it with a layer of sand.
- Wait until foliage dries up, and then pull it off. Don't braid or tie the foliage, which cuts off air circulation and reduces the amount of sun that reaches the leaves.
- Disguise dying bulb foliage by planting with shallow-rooted annuals. Use tall, lacy-leafed annuals that won't block the light, such as cosmos or baby's breath or love-in-a-mist, which is tall and branching, comes quickly into bloom in spring, and dries up in summer. Avoid plants that need constant watering, which is bad for dormant flower bulbs.

***BUTTERFLIES, to attract to the garden***

- Cultivate flowers that are mauve, blue, deep pink, crimson, or deep yellow. Researchers have discovered that butterflies prefer flowers of these colors.
- See Internet Resources, page 153 for a list of organizations that offer suggestions for attracting butterflies to the garden.

# C

**_CABBAGE, hints on_**

- Plant cabbages among taller plants that provide shade. This can also protect cabbages from flea beetles, which prefer to feed in full sun. Or plant them under lightweight floating row covers, which can stay on all season.
- Plant fast-maturing varieties to avoid cabbage worms. Planting with celery will also repel cabbage worms.
- Avoid planting cabbage with tomatoes, kohlrabi, or pole beans, which are not beneficial companions.
- Harvest a second head when the first has reached maturity by cutting the cabbage but leaving the stem and large outer leaves intact. When tiny new heads form, remove all but one so the remaining head has room to grow. Or leave the tiny heads in place to produce miniature cabbages.
- Obtain four new smaller cabbages after the main head has been harvested by slitting the remaining stalk into quarters about one and a half inches deep.
- Delay harvesting for a few more weeks by pulling up the cabbage slightly, twisting the head a half turn, and then letting it settle back. This will break off some of the feeder roots and keep it from going to seed or splitting.
- Pick off cabbage loopers and cabbage worms, and drop them into a bucket of soapy water. Look for light-green eggs and caterpillars on the undersides of the leaves. Alternatively, dust the cabbages with a fine spray of flour while the dew is still on the leaves. Wash it off in two days with a fine spray from the garden hose. Repeat if necessary.

**_CABBAGE, CHINESE, hints on_**

- Blanch the cabbages for milder flavor by placing half-gallon milk cartons over the plants when they are eight to ten inches tall. Open both ends of the cartons, and keep them on the cabbages for a couple of days.

- Use the young leaves for salads and the mature leaves for sautéing and stir-frying.

***CABBAGE PESTS, to control***

- Protect cabbage from whiteflies with sections of old pantyhose. Slip the hose over young cabbage heads, knot the top, and pull the hose down to soil level. The nylon will expand as the plants grow. Basil and sage planted between rows will also repel whiteflies.
- Provide protection against root flies by placing a six- to eight-inch square of tar paper, cardboard, or heavy-duty foil around the base of the young plant. Sow the seedlings through an X-shaped slit in the center of the square. Anchor the paper with a little soil, and leave it in place all season. Growing thyme nearby will also repel cabbage-root flies.
- Repel cabbage loopers by interplanting garlic, onion, rosemary, or sage.
- Plant nasturtiums as a lure and trap. The caterpillars love nasturtium leaves and might dine on them rather than the cabbages.

***CALCIUM SOLUTION, to make***

- Soak clean, crushed eggshells in water for three or four days to leech out the calcium, and then use the water on plants to help strengthen plant roots. Soak the eggshells outdoors to avoid the odor.

***CAMELLIAS, hints on***

- Avoid fertilizing camellias while buds are forming. Camellias are dormant while blooming, and fertilizing upsets the resting period.
- Transplant or set out new plants when they are in flower. This is when they are in their most dormant period.
- Spray the foliage with a fine mist during dry, hot spells. Do this along with regular watering. Camellias appreciate the overhead watering during hot weather.

- Lessen flower drop by removing all but one or two flower buds from each branch-end cluster, and remove all flower buds except for a single bud for each two to four inches of branch. This also makes for a more dramatic showing.

***CARROTS, hints on***

- Avoid planting celery or dill with carrots; they are not beneficial companions.
- Hasten germination by chilling the seeds in the refrigerator for a day or two, then soaking them for an hour in warm (not hot) water before planting.
- Mix seeds with coarse sand to help plant evenly. Or mix the seeds with some fresh coffee grounds before sowing them. The coffee aroma repels root maggots and other pests, plus the extra bulk makes the seeds easier to sow.
- Avoid deformed roots in heavy soils by planting the shorter varieties.
- Sow the carrots with radishes or lettuces to mark the rows. When harvesting the radishes or lettuces, you will aerate the soil.
- Prevent a crust from forming on the soil by sprinkling with a fine spray, keeping the area shaded, or covering the seeds with damp newspaper until germinated.
- Protect plants from carrot rust flies by covering them with a polyester row cover, mosquito netting, or fine cheesecloth. You can leave the protection on all season if desired. Another way to protect the plants is to interplant the carrots with onions or leeks, which confuse the carrot flies' sense of smell.
- Trap carrot rust flies with a quart-size milk carton coated with petroleum jelly. Place it near the plants, and wipe insects off with a putty knife.
- Thin to remove excess seedlings when the carrots' tops are about one to two inches tall. Steam the tiny carrots, or chop the entire plants, tops and all, and add to tossed salads.

- Harvest by hand pulling. Digging them up with a spade or a fork could injure the carrots.

### *CATS, to deter from garden beds*

- Scatter pieces of grapefruit, lemon, or orange peel in the areas you want the cats to avoid or around plants they favor. Cats dislike the smell of citrus.
- Place plastic netting or chicken wire over a newly prepared area to prevent cats from using it as a litter box. Or lay chicken wire outside of the garden beds the cats are favoring.
- Place a few vinegar-soaked rags in the beds.
- Use cocoa husks or shells as a mulch in planted areas. Cats don't like the smell and will avoid areas where cocoa hulls are used. (Rake them around periodically to keep them fresh.)

### *CAULIFLOWER, hints on*

- Avoid planting with tomatoes, which are not beneficial companions.
- Protect the plants from egg-laying insects by slipping a piece of pantyhose over the cauliflower when it starts to form a head; it holds the leaves snugly and stretches as the head grows.
- Cover the cauliflower when the heads are the size of a half-dollar by gathering the leaves together and securing them with a rubber band. Doing this prevents the cauliflower from producing chlorophyll and turning green. The other option is to choose a self-blanching type, whose inner leaves curl over the top and do the job for you.
- Foliar feed with liquid seaweed extract if the heads turn brown or the leaf tips die back. This condition is due to a deficiency of boron in the soil.

### *CELERY, hints on*

- Avoid planting celery with carrots, parsnips, or potatoes, which have an adverse effect on the growth of celery.
- Start plants indoors; celery seeds are slow to germinate. Or soak seeds overnight, and then mix them with dry sand before planting in

the garden. Keep the seedbed covered with moist burlap or newspaper until seedlings appear.

- Plant the celery in a circle so that the plants can shade each other.
- Blanch the celery stalks when the plants are twelve to fifteen inches tall. Wrap several layers of newspaper or cut-up brown-paper grocery bags around the stalks, leaving the top leafy growth uncovered, and then tie with string. Or prop topless and bottomless cardboard milk cartons over the celery. Continue to water as usual. The celery should be ready in two to three weeks.
- Grow self-blanching celery in blocks instead of rows; a certain amount of blanching improves the celery.
- Treat distorted leaves or cracked stems by spraying the celery with seaweed extract every two weeks until symptoms disappear.

***CHERRIES, hints on***

- Prevent birds from eating the fruit by covering the tree with plastic netting. Tie the bottom of the netting to the trunk so the birds can't get underneath. Or plant yellow cherries; the birds don't go for the yellow varieties as much.
- Surround the tree with a dense ground cover or mulch to keep fruit fly larvae out of the soil. If larvae cannot reach the soil, they die.

***CHRISTMAS TREES, hints on***

- Choose a fir tree for fragrance, or pick a fir or Scotch pine if you want a tree that will last longer.
- Save the tree after Christmas, and place it in a secluded spot in the garden for a wildlife shelter and bird feeder. When spring arrives, use the tree as a pea trellis. Anchor it in place, then plant the peas around the outside.
- Recycle the tree after Christmas. First, strip off the needles, and add them to the compost pile, or use them as mulch around acid-loving

plants or shrubs. Next, cut off all the side branches, and use them for winter protection of garden plants or as mulch for acid-loving plants. Finally, use the trunk for firewood. (Let the wood dry out a season or two before using.)

***CHRISTMAS TREES, LIVING, hints on***

- Dig a hole for the tree before the ground freezes if you live in a cold-winter area. Fill it with loose straw, and then cover the hole with plywood. Save the soil in an area where it's protected from freezing so you can use it later when planting the tree.
- Choose a species that grows well in your climate. Many evergreen trees cannot thrive in warm winters and may gradually decline.

***CHRYSANTHEMUMS, hints on***

- Obtain a bushy plant with lots of flowers by pinching off the growing tips as soon as the plant is six inches tall; then keep pinching the tips off every three or four weeks until the plant is about twelve to eighteen inches tall (or until buds start forming).
- Choose cushion mums, which are low and shrub-like, if you want to avoid the pinching process.
- Maintain a constant moisture level, but let the ground dry out between each watering. Too much water will cause the plants to rot; too little will cause the plants to get woody.
- Keep the colors more vivid in very hot climates by providing afternoon shade.
- Avoid nematode problems by not growing chrysanthemums in the same place for more than a few years.
- Start a new plant by bending a low branch to the ground and covering a section of it with soil. Make a nick in the part that touches the ground, and weigh it down with a rock. When the buried section has developed roots, cut it off from the parent plant.

***CITUS, hints on***

- Keep the trees clean. Beneficial insects can't reach the harmful ones if they are hidden beneath a layer of dust. Spray the tree when dusty with a solution of one to two teaspoons of mild liquid dish soap per gallon of water.
- Keep ants off trees with a barrier. Wrap a strip of plastic wrap tightly around the trunk, and then apply petroleum jelly, heavy-duty oil, or Tanglefoot, to the center of the plastic wrap. Renew the coating frequently to keep it sticky, and remove it when no longer required.
- Germinate fresh citrus seeds the easy way by putting them in with a potted plant you keep on the moist side. Bury the seeds one inch below the soil surface, and repot them when they germinate.

***COMPANION PLANTING, hints on***

- Plant basil near tomatoes to repel hornworms and flies. Chop and scatter basil leaves in vegetable and flower beds to repel aphids, mosquitoes, and mites.
- Plant catnip by eggplant to deter flea beetles and near cabbage to deter cabbage pests.
- Plant dill to lure pests away from early squash and to trap pickleworms before planting late melons. Pull and destroy trap plants as soon as they are infested.
- Plant French marigolds among vegetables, either as a trap for stink bugs or as a repellent for tomato hornworms, squash bugs, Mexican bean beetles, cucumber beetles, whiteflies, and nematodes. Dig the marigolds into the soil at the end of the season to help control nematodes.
- Plant garlic near cabbage, corn, carrots, lettuce, peas, and tomatoes to repel many pests. Also plant garlic with roses to repel aphids and Japanese beetles.
- Plant horseradish by potatoes to repel potato bugs.
- Plant mint by broccoli, Brussels sprouts, and cabbage to repel the egg-laying whitefly, with radishes to discourage flea beetles, and with tomatoes to repel tomato hornworms.

- Plant nasturtiums around squash to repel squash bugs and cucumber beetles and with cabbages as a lure for cabbage worms and loopers. Nasturtiums also repel aphids.
- Plant onions next to beets, cabbages, and carrots to keep bugs away. Onions also repel cutworms.
- Plant radishes around cucumbers and squash to repel cucumber beetles.
- Plant tansies next to cucumbers, pumpkins, melons, and squash to repel squash bugs.
- Crush a leaf of the companion plant now and again in order to release the aroma.

***COMPOST ACTIVATOR (material added to the compost to speed bio-degradation in new piles)***

- Ask for a couple of handfuls of compost from the center of a neighbor's pile to add to your pile.
- Add a little natural activator using what you have available: powdered meal (alfalfa, blood, bone, cottonseed, crab, feather, hoof, kelp, soybean), seaweed, coffee grounds, fish emulsion, rich organic soil, fresh or dry manure (cow, steer, sheep, goat, llama, poultry, rabbit, hamster, guinea pig), or guano.

***COMPOST AERATOR (tool for stirring and turning the compost and introducing oxygen into the bin, which is necessary for helping the material break down and keeping the compost active)***

- Use a reaching pole hook, metal bar, old broom handle, garden fork, or shovel.

***COMPOST BIN, alternative for***

- Dig a ten- to twelve-inch-deep trench in an unused part of the garden, away from growing plants, and put kitchen scraps into it: vegetable and fruit peelings and scraps, coffee grounds, paper tea bags, rinsed and crushed eggshells, plus any garden refuse, excluding meat, fat, and bones. Add a little soil to the trench each time you put material into it to keep it sanitary. When the trench is full, start another one. You can also chop or puree kitchen scraps in the blender with water. Pour the mixture into the trench or into small holes next to tomatoes, peppers, and other plants, and cover the mixture with soil.

***COMPOST MATERIALS***

- Use the following items to make compost: fruit and vegetable scraps, coffee grounds, tea leaves, crushed eggshells, shredded leaves, pine needles (although they take a while to break down), grass clippings, hay, shredded paper and newspaper (no color or glossy ads), dust from the vacuum cleaner, plus manure, feathers, sawdust, and mushroom compost (see *COMPOST MATERIALS, to acquire*). Do not include dog or cat feces, weeds (unless the compost pile has heated up sufficiently), fat, or meat scraps.

***COMPOST MATERIALS, to acquire***

- Manure: riding stables, farms or ranches (sheep, goat, hog, dairy, or egg), racetracks, zoos, and petting zoos
- Feathers: poultry farms
- Sawdust: lumberyards, firewood lots, cabinetmakers, and high-school woodshops
- Mushroom compost: mushroom growers
- Coffee grounds: coffeehouses, hospital cafeterias, and restaurants
- Spent hops: local breweries or home brewers
- Grass clippings and leaves: local schools, lawn-maintenance services, and neighbors
- Ground-up leaves and bark: utility companies and tree-care or tree-removal companies
  1. Call the county or municipal Department of Public Works to inquire whether free compost is available in your area.
  2. Search online for free mulch in your city or town.

***COMPOSTING, quick tricks for***

- Speed up the breakdown of the compost material, and keep the moisture level optimum by covering the compost container with clear heavy plastic. (Don't let it touch the compost.) The plastic helps retain heat while still allowing sunlight to warm the heap. It also prevents evaporation and protects the compost from heavy rainfall, which might add too much water.
- Put all the compost materials in a heavy-duty garbage bag, add a shovelful of soil and one-quarter cup of all-purpose fertilizer, such as 10-10-10, and then moisten the mixture thoroughly. Seal the bag, and set it in a sunny area. Roll it over carefully twice a week for two months.
- Make compost in three or four weeks with crumbled or shredded leaves and grass clippings. Fill a large clear-plastic bag almost full with the leaves and clippings. Mix in a five-pound bag of high-nitrogen fertilizer, and then add enough water to moisten the leaves and grass. Put the bag in a sunny spot, and turn it over three or four times a week.

***CONTAINERS FOR PLANTS, hints on;*** see also *BUCKETS, hints on*

- Use small garbage cans, five-gallon drums, whiskey or wine barrels, buckets, dishpans, wastepaper baskets, hard-shelled luggage, joint-compound or industrial-paint buckets, frosting or pie-filling pails from bakeries and doughnut shops, five-pound coffee cans, large old cooking pots, wooden boxes, reusable plastic shopping bags, or old barbecues and their lids. Drill or poke drainage holes along the lower edges of the containers near or in their bottoms. Line metal containers with bubble wrap to insulate them, taking care not to block the drainage holes.
- Inquire at produce markets for wooden crates, fruit boxes, and bushel baskets. Line them with doubled-up burlap before adding the potting soil.
- Check the farm-and-garden section on Craiglist.org for low-cost or free pots and planters.
- Line an old plastic laundry basket with a heavy-duty plastic garbage bag, and poke a few holes in the bottom of the bag (plus the laundry basket, if it's the solid kind).

- Put an old wheelbarrow into service. Make enough holes in the bottom for drainage.
- Line the bottoms of containers resting on bare soil with wire mesh. If you don't line the pots or containers with mesh, move them occasionally so you can check for and remove slugs and sow bugs. Check also that the roots don't grow through the drainage holes into the soil.

***CONTAINERS FOR PLANTS, HEAVY PLASTIC, to make drainage holes in***

- Hold the container upside down over a candle until the plastic softens; then lay the container on a board right side up, and use a hammer and chisel to make a hole in the softened plastic.
- Drill a hole with an electric drill at low speed using a quarter-inch bit.

***CONTAINERS FOR PLANTS, METAL, to make drainage holes in***

- Freeze water in the container; then turn the container upside down, and poke holes in the bottom with a nail and hammer.
- Set the container on a piece of wood (right side up) and hammer holes in the bottom with a large nail or a Philips-head screwdriver.

***CONTAINERS FOR PLANTS, THIN PLASTIC, to make drainage holes in***

- Heat the point of a nail, a metal knitting needle, or a straightened-out wire coat hanger over the flame of a gas stove or candle, and use it to melt holes in the bottom of the container.

***CONTAINERS FOR SEEDS STARTED INDOORS;*** see *SEEDS STARTED INDOORS, containers for*

***CORN, hints on***

- Improve the odds of corn germination by soaking the seeds in room-temperature water for twelve hours prior to planting.
- Interplant with pole beans. The beans provide nitrogen for the corn, and the corn can provide a trellis for the beans. Squash and cucumber are

also beneficial companions. Squash and cucumber provide mulch for the corn, which likes its roots cool and its tops in the sun.

- Avoid planting with tomatoes. The tomato worm and corn earworm are identical.
- Plant corn in blocks of at least four per row. Or plant six to seven seeds in hills of mounded soil about three feet apart; then thin to the three strongest plants in each hill. This method not only saves space in the garden but promotes pollination.
- Bury a fish or fish scraps in each block or circle planted. Fish is rich in phosphorus and potassium, as well as iron. (Ask a fishmonger or a supermarket manager of the meat-and-fish department for any fish scraps—heads, tails, etc.)
- Plant the seeds in a three- to six-inch-deep trench during hot dry weather, and then fill in the trench as the plants grow. Seeds planted this way makes for sturdier corn.
- Select corn that matures at different times if you grow more than one variety, or stagger the planting times. This will not only prevent cross-pollinating but extend the harvest.
- Put down strips of aluminum foil as a mulch. In tests, it increased the yield substantially. Lay the foil over newspaper or plastic after the corn is up.
- Leave the stalk suckers alone. Removing them might cause a setback to the plant or reduce the crop.
- Avoid overhead watering. Doing so can wash pollen off the tassels.
- Harvest late in the afternoon. Sugar production occurs during sunlight hours.

***CORN, to protect from birds***

- Cover the seeds with a very light straw mulch, or start the seeds indoors in peat pots.
- Keep crows from eating the sprouting seeds by planting potatoes first. Plant the rows about three feet apart, and when the potatoes are about a foot tall, plant corn between the rows. The foliage scares the crows.

- Put a berry basket (the kind that strawberries and cherry tomatoes come in) over each planted seed. Anchor them if necessary, and then remove the baskets when the sprouts reach the tops of the baskets.
- Crisscross string all over the patch, using wooden stakes, pieces of branches, or chopsticks to hold the string.
- Plant the seeds in a shallow trench, and cover it with chicken wire. Or make a shallow tent with the wire, and place it over the planting area. Remove it when the sprouts are well established.
- Cover the ears with paper bags after the silk turns golden brown.

***CORN, to protect from raccoons***

- Cover each ear with a plastic bag when the ears are just about ripe enough to pick. Hold the bag in place with a rubber band.
- Plant winter squash or pumpkins around the perimeter of the corn early in the season so they'll carpet the ground when the corn is ready. (Guide the squash and pumpkin runners in among the corn when they start growing.) Raccoons don't like to walk over the big prickly leaves.

***CORN PESTS, to control***

- Plant susceptible crops as early as possible to avoid corn earworms.
- Put clothespins over the silk channels when the silks first emerge, or use rubber bands.
- Apply several drops of mineral oil to each stalk at the tips of the ears about three to seven days after silks appear. Apply three times at five- to six-day intervals. Use a medicine dropper, plastic detergent squeeze bottle, or pump-type oil can. Remove any pests after the silk has browned.

***CUCUMBERS, hints on***

- Avoid planting cucumbers near melons, pumpkins, or squash or in the same area they were planted the previous year. They are all kin and prone to the same pests and diseases.

- Train cucumbers on a six-foot trellis (see *TRELLIS, hints on*) or plant with corn, which will provide support for the cucumber stems. Growing vertically makes cucumbers easier to pick and produces fruit that is long and straight, without kinks or curls. Research has also shown that cucumbers grown vertically will produce twice as much good fruit and have fewer diseases than cucumbers grown on the ground, and they will produce fruit up to five weeks longer than the same variety grown without trellising. Stop their growth when they reach the top of the trellis by pinching off the fuzzy ends. This way the plants can devote energy to fruit production rather than more foliage.
- Protect young plants from cucumber beetles with cheesecloth, row covers, or nylon netting. Remove the covering before blossoms set so pollinating insects can pollinate the blossoms (unless they are all-female varieties).
- Trap cucumber beetles with shallow tins of water to which a little cooking oil has been added. What also works is planting radishes as a lure for the beetles. Leave the radishes in place all season for full effectiveness, and then destroy the radishes.
- Move the vines as little as necessary, which can destroy blossoms, drive away pollinating insects, and increase the number of misshapen fruit. Also avoid handling or brushing up against the foliage when it's wet to keep from spreading disease.
- Use warm (not cold) water to water cucumbers. Set out a bucket of water to warm up in the sun. Or if the hose has been out in the sun, water the cucumbers first. Cold water shocks the leaves when the sun is hot, plus warm water speeds up the growth of all crops.
- Avoid watering cucumbers overhead, since they are susceptible to mildew. Provide a constant supply of moisture by one of these methods:
  1. Punch holes in any large-size coffee or juice can and sink it in the soil before you plant the seeds. Leave one inch of the can above the soil

level, and fill the can with water whenever you water the cucumbers. If you want, you can put a flat rock on top of the can to prevent evaporation.

2. Drive one-and-one-half to two-inch-wide PVC pipes into the ground before sowing the seeds. When the plants are established, pour water into the pipes so it will be directed closer to the roots.
3. Place a soil-soaker hose next to the cumbers, and leave it in place all season; then connect it to the main hose when watering.

- Harvest the cucumbers daily to lengthen the life of the plant and increase the size of the crop.

***CUSHION FOR GARDENING;*** see *KNEELING PAD, to make*

***CUTWORM COLLARS (cylinders to protect seedling stems from cutworms)*** *to make*

- Split fat plastic drinking straws up one side, and then cut them into one-and-a-half- to two-inch sections.
- Use Styrofoam, paper, or yogurt cups. Cut the bottom sections off the cups to make cylinders.
- Cut cardboard tubes from toilet paper, wrapping paper, or paper towels into two-inch lengths.
- Cut both ends off frozen-juice containers or canned-food containers to make collars. (Wash them well before using.)
- Make aluminum-foil collars from two-inch-wide strips of foil, cut into four- to five-inch lengths, folded in half, and then folded in half again.
- Use milk cartons with the tops and bottoms cut off to encircle large transplant stems. Or cut the cartons in half.

***CUTWORMS, to control;*** see *SEEDLINGS IN THE GARDEN, to protect from cutworms*

# D

### ***DAHLIAS, hints on***

- Plant red and vivid-colored dahlias where they will get afternoon shade; otherwise the sun will bleach out the color.
- Obtain extra-large blooms by pinching off all side buds and leaving the centers intact.
- Have lots of blooms by pinching off the growing tips when the plant is six to ten inches tall and again after it grows another six inches. This will promote branching and more flowers. Small flowering dahlias need only the first pinching.
- Take two- to four-inch cuttings of nonflowering young growth in the spring. Plant in rich, loose soil where they are to grow, and water well until established.

### ***DELPHINIUMS, hints on***

- Add polymer granules to the planting holes to increase the soil's water-holding capacity. These nontoxic granules absorb up to ninety times their weight in water and gradually release it into the soil. Delphiniums like more water than most flowers, and with polymers you can water them without overwatering nearby plants.
- Put in six-foot stakes when planting delphiniums, and then stake them as they grow. Or tie them close to the summit of the spike after they have attained full growth.
- Avoid getting water on the leaves, as delphiniums are prone to fungal disease.
- Cut back the flower spike to eight inches right after the first bloom, leaving plenty of foliage at the base, and then give them a dose of fertilizer. When the new shoots are several inches high, cut the old stalks to

the ground. Weather permitting, it's often possible to get a third round of blooms. Make sure you cut back the stems as soon as the blooms start to fade and before they set seed.

- Choose heat-tolerant delphiniums if you live in a warm climate. They are easier to grow.

***DILL, hints on***

- Avoid planting dill with carrots or tomatoes. Dill has an allopathic effect, which suppresses the growth of carrots and tomatoes.
- Plant dill against a wall or fence or toward the back of the flower bed.
- Make one sowing early in the season and another later on. While the first sowing is going to seed, the second will be on its way.
- Let a few plants bloom to provide food for the beneficial insects that favor this member of the carrot clan: wasps, bees, microscopic tachinids, and ichneumon wasps.
- Keep a plant on a sunny windowsill. Dill is one of the herbs that can happily exist indoors.

***DISEASES, PLANT, hints on***

- Buy disease-resistant species whenever possible and especially if you have encountered problems in the past. The resistance qualities are usually listed on the seed package right after the name of the variety. For instance, a *v* and an *f* indicate that the plant is resistant to verticillium and fusarium wilt.
- Provide optimum conditions for plants that are prone to mildew, such as lilies, phlox, roses, and zinnias. Make sure they are in a sunny location with good air circulation. Thin out extra stems if necessary to ensure air movement.
- Wash the plants with a solution of one teaspoon of mild liquid dish soap per gallon of water. Spray first thing in the morning. Soap has antiseptic qualities, which is useful against many plant diseases.

- Spray mildly diseased plants with liquid seaweed emulsion diluted to half the rate recommended on the label, and spray once a week. The micro-nutrients found in kelp act like a vitamin shot by encouraging better growth and development. Compost tea also contains beneficial organisms, nutrients, and many other compounds. Research shows that spraying with compost tea or liquid seaweed emulsion can prevent or reduce certain disease problems. Add a few drops of mild liquid dish soap to help the solution stick to the plant leaves.
- Spray mildew or other fungal diseases with a fungicide (see *FUNGICIDE SPRAY, to make*).
- Discard any seriously diseased plants before the disease spreads to other plants. And don't grow another plant of the same species in the same spot; the new plant may become afflicted with the same disease.
- Wash your hands after touching a diseased plant, and sanitize your tools by cleaning them with rubbing alcohol or a disinfectant wipe and letting them air dry.

### *DROUGHT-TOLERANT PLANTS*

- Use plants native to your area. They are rugged and require less water.
- Plant ornamentals: agastache, agave, blanket flower, bougainvillea, cactus, catmint, coreopsis, desert rose, euphorbia, globe thistle, golden marguerite, lantana, lavender cotton, lavender, lithops, oleander, poppy, portulaca, purple coneflower, rock daisy, Russian sage, salvia, sedum, soapwort, Spanish lavender, verbena, wallflower, and yarrow.
- Go to Google and type in "Nifty 50 Plants for Watersmart Landscapes" for a downloadable PFD made available through the San Diego County Water Authority at watersmartsd.org.

# E

### *EARWIGS, to control*

- Dampen newspaper, roll it into a tight cylinder, and then place it where earwigs appear. It should be full of earwigs the next morning. Empty the earwigs into a container of soapy water. Continue until no more earwigs appear. Corrugated cardboard, a few pieces of old garden hose tied together, or a loosely rolled cardboard cylinder will also work.
- Grease the top inside third of a tin can with bacon grease, and fill it with water up to the grease. Lay a piece of brown paper over the can, and put it out where the earwigs appear. Dispose of the can when it is filled with earwigs.
- Cut the top off a milk carton, and fill the bottom with one cup of water and one teaspoon of vegetable oil. Put it out where the earwigs appear; then dispose of it when it's full.

### *EGGPLANT, hints on*

- Get a jump start by buying eggplants as seedlings from the nursery. The seeds take a long time to germinate.
- Make sure the plants get the heat they need in these ways:
  1. Preheat the soil with clear plastic.
  2. Plant eggplant in the sunniest spot available or against a wall that will reflect the sun.
  3. Put hot caps over the plants to keep them warm (see *HOT CAPS, hints on*).
  4. Grow a plant in a five-gallon container so you can keep it in the sunniest part of the garden or patio and move it around.

- Protect the plants from flea beetles by covering young plants with cheesecloth, mosquito netting, or row covers. Remove the covering before blossoms set so pollinating insects can pollinate the blossoms. Another option is to place the plants in tomato cages. Wrap the cages with netting or row-covering material, then fold over the edges at the top and sides, and staple them closed. Remove the covering for pollination when plants begin to blossom.
- Control flea beetles by planting parsley, dill, and Queen Anne's lace to attract predatory beneficial insects and give them a base of operations.
- Help the plants with a dose of magnesium by
    1. mixing one teaspoon of Epsom salt with the soil at the bottom of the hole when transplanting;
    2. spraying the plant when blossoms appear with a solution of one tablespoon of Epsom salt dissolved in one quart of water (pour the remainder in a ring around the plant at the drip lines); or
    3. adding one-quarter cup of Epsom salt to the soil around each plant when blossoms appear. Epsom salt consists of magnesium sulfate, which helps set blossoms and improves the soil, and sulfur, which eggplants need.
- Pinch off the tops when the plants are young to promote bushy growth.
- Prevent too much fruit setting by pinching off all but four to six blossoms per plant. Or if you enjoy tiny eggplants, allow plants to produce freely.
- Hold back on watering in the fruit-setting stage, and feed plants with wood ashes or seaweed extract.
- Extend the harvest period by picking the eggplants whenever the skins appear glossy and they are large enough to use, around three to five inches, depending on the variety. Eggplants taste best when young and undersized—they're more tender and of better quality than those left to grow to full size. They taste bitter when overripe.
- Harvest the eggplants using a sharp knife or pruning shears, and wear long sleeves and thick gloves when harvesting or working around the plants. The eggplant stem has prickles, which can irritate the skin.

### *EGGSHELLS, uses for*

- Add crushed, baked eggshells to potting or garden soil. Eggshells contain around 93 percent calcium carbonate (which is a good source of lime for acidic soils), plus nitrogen, and trace amounts of potassium and phosphorus.
- Sprinkle the dried shells around flowers and vegetables to keep slugs away. Bake the eggshells in a 250°F oven until they turn brown and crisp; then crush them in a plastic bag with a rolling pin or whirl them in the blender or spice grinder.
- Soak eggshells in water for several days to leech out the calcium (do this outdoors to avoid the odor), and then use the water on plants to help strengthen their roots.

### *ENDIVE (ESCAROLE), hints on*

- Blanch endive (a.k.a. escarole) to lessen its bitterness and improve its flavor by tying the outer leaves up over the center leaves. Use strips of nylon pantyhose, string, or rubber bands. Do this when the plants are six to eight inches across (about two or three weeks before harvest) and when the leaves are dry. If you only have a couple of plants, cover each whole plant with a cardboard box or a large plant pot with the drainage holes covered.
- Harvest the endives by taking a few outer leaves at a time, or cut out the center tender leaves. New center shoots will quickly develop for another harvest. Endive is a good lettuce substitute and is more cold and heat resistant than lettuce.

### *EVERGREEN BOUGHS (FIR, PINE, SPRUCE), to keep fresh*

- Recut the ends of evergreen boughs and pound them with a hammer so they will absorb more water; then soak them overnight.
- Soak the cut ends of evergreen boughs for twenty-four hours in a solution of one-half cup of brown sugar to one gallon of water. Dissolve the sugar in a small amount of hot water, and then pour the solution into a

container of lukewarm water. Make a fresh cut if the boughs aren't put in the sugar water immediately; otherwise the sap will seal over the cut end and prevent water uptake. After conditioning the boughs, remove them from the water to use as decoration, or, if displaying them in vases, just add fresh water to the sugar solution as it evaporates.

# F

***FERTILIZER, hints on***

- Enrich the soil with kitchen scraps and food past its prime (no meat or fat though). Add the kitchen scraps to the compost, or bury them in the garden ten inches deep.
- Use compost as a general all-purpose fertilizer: spread it two inches thick over the entire garden each spring.
- Use banana peels. The peels provide 41.76 percent potash and 3.25-plus percent calcium, magnesium, sulfur, phosphorus, sodium, and silica. The peels are especially beneficial for roses and other flowers, tomatoes, peppers, and eggplants. Chop the peels into small pieces, and place them just below the surface of the soil. They rot quickly.
- Keep citrus skins; they are rich in nitrogen. Bury them in the garden, or add them to the compost heap. Don't go overboard, though; earthworms are not that keen on them.
- Save fish heads, tails, and bones, and bury them in the garden. Fish is a good source of phosphorus and potassium, plus iron. To keep animals from digging them up, bury the fish parts ten inches deep.
- Give earthworms a treat with used coffee grounds; they love them. Coffee grounds contain 4 percent nitrogen, 2.2 percent phosphorus, and 1.19 percent potassium, and they will not only enrich the soil but also check the growth of fungus. Use especially around acid-loving plants.
- Save all the ashes from a wood-burning stove or fireplace. Wood ashes contain 3 to 7 percent potassium and 15 percent phosphorus and are an excellent source of potash; however, wood ashes contain 20 to 50 percent lime and can raise the pH of the soil. Don't use an abundance of ashes if the pH of the soil is above 7.0 or where acid-loving crops will grow. Also avoid using them where potatoes will grow.

- Use cooled water from unsalted boiled potatoes and other vegetables in the garden. The water has nutrients such as nitrogen, phosphorus, and potassium, plus other trace minerals that are beneficial to plants. Also be sure to save the water from rinsing sprouts—another source of nutrients.
- Use the water from your aquarium to water vegetable plants or roses; it's loaded with minerals.
- Collect fresh seaweed washed up on the beach. Take a little here and there and only part of each plant. Never take the whole plant. Seaweed is an excellent source of potassium (or potash). Rinse it to remove the sea salt; then use it as a mulch, or chop it into small pieces and incorporate it into the soil. Kelp provides some seventy trace elements that plants need in very small quantities. It also contains growth-promoting hormones and enzymes. (See also *SEAWEED TEA, to make*)
- Leave the grass clippings on the lawn after mowing to provide nitrogen.
- Make compost tea by filling a burlap bag or old pillowcase with compost and suspending it in a large container of water for four or five days. Dilute as needed until the water is the color of weak tea, and then pour it around the plants.
- Use blood meal, fish emulsion, alfalfa meal or pellets, or soybean meal for nitrogen (N).
- Use bone meal, fish bone meal, or rock dust/soft rock phosphate for phosphorus (P).
- Use glauconite (greensand), kelp meal, or granite dust for potassium (K).

***FERTILIZING, hints on***

- Mix liquid fertilizer in a washed-out gallon-size bleach or detergent bottle to make measuring the water easier.
- Put bagged fertilizer into an empty shaker container with holes, and then shake the fertilizer over the soil for more even distribution.
- Choose slow-release organic fertilizers to build the soil.

- Resist fertilizing a stressed, sick, or diseased plant, don't fertilize a new transplant until it's established, and don't feed container or bedding plants you just purchased; they were probably well fed at the nursery.
- Apply liquid fertilizer to damp soil, never to dry soil.
- Keep all fertilizer away from seeds, stems, and root hairs of young plants; it can burn.
- Mix some flour with dry fertilizer before you spread it on the lawn. It will indicate what areas have been covered and will water in without harming the grass.
- Add one-quarter teaspoon of mild liquid dish soap or vegetable oil per gallon of solution when foliar feeding plants. Doing so will help the solution stick to the leaves and ensure better coverage. Spray until the liquid drips off the leaves, paying special attention to the undersides of the leaves. Spray in the early morning, and never spray when it is windy.
- Spray plants with seaweed emulsion when buds are forming on fruits, flowers, and vegetables; and when root crops are beginning to form mature roots. Give the fruit another dose when tiny fruits begin to form. Spray greens with a diluted solution on a biweekly basis to encourage rapid growth.

### ***FLEA BEETLES, to control***

- Plant susceptible crops two weeks later than normal to avoid peak flea beetle populations.
- Plant mint and catnip as a repellent or radishes as a trap crop.
- Put out yellow sticky traps (see *STICKY INSECT TRAPS, to make*).
- Place diatomaceous earth around the plants.

### ***FLOWERS, best for attracting beneficial insects to the garden***

- The best flowers for attracting beneficial insects to the garden are asters, baby blue eyes, baby's breath, black-eyed Susan, candytuft, coreopsis, cornflower, goldenrod, marigold, nasturtium, small chrysanthemum, strawflower, sunflower, sweet alyssum, white evening primrose, and annual wildflowers.

***FLOWERS, best for cut arrangements***

- The best flowers for cut arrangements are annual or China aster, baby's breath, bells of Ireland, calendula, carnation, cockscomb, cornflower, cosmos, delphinium, dianthus, marigold, nasturtium, petunia, scabiosa, snapdragon, stock, sweet pea, and zinnia.

***FLOWERS, best for drying***

- The best flowers for drying are acrolinium, bells of Ireland, cockscomb, globe amaranth, honesty, love-in-a-mist, pearly everlasting, pink paper daisy, statice, strawflower, and winged everlasting.

***FLOWERS, best for fragrance***

- The best flowers for fragrance are candytuft, carnation, cleome, forget-me-not, four-o'clock, freesia, mignonette, moonflower, nasturtium, petunia, Siberian wallflower, stock, sweet alyssum, sweet pea, sweet William, tall garden phlox, tuberose, and verbena.

***FLOWERS, best for window boxes***

- The best flowers for window boxes are ageratum, candytuft, coleus, lantana, lobelia, dwarf marigold, dwarf nasturtium, pansy, petunia, dwarf phlox, salvia, snapdragons, sweet pea, torenia, verbena, vinca, and dwarf zinnia. Pinch off the tops of snapdragons and petunias after planting so they will grow bushier and produce more blooms.

***FLOWERS, to dry***

- Gather the flowers on a warm, dry day right before their prime, when they are not quite fully opened but have the longest stems possible.

- Air-dry flowers by hanging them upside down in bunches of three or four in a cool, dark, airy place until fully dry. Strip off the leaves before drying.
- Dry flowers in silica gel, clumping cat litter, borax, or a mixture of three cups of borax to one cup of cornmeal. Place a layer of the mixture in a cardboard shoebox, and then put the flowers on top, facing downward. Cover the flowers gently with more of the drying medium, put the lid on, and then seal the lid shut with cellophane tape. Set the box in a warm, dry place for five days to two weeks, or until dry. Remove the dried flowers gently, and brush off any excess drying medium with a soft paintbrush.
- Dry flowers in the microwave. Wrap them loosely in a paper towel or napkin, and then microwave them for two minutes on High. If the flowers are dry and crisp, they are done. If not, microwave them for another minute or so. Dry leaves in the same way. Place some inside a folded paper towel, and weigh the paper towel down with a microwave-safe container to prevent the leaves from curling.

***FLOWERS, CUT, hints on***

- Make flowers with short stems appear taller by inserting the stems into clear-plastic drinking straws before adding them to the arrangement.
- Keep a damaged stem erect by taping a toothpick to the area to act as a splint.
- Prop up a drooping flower head by inserting toothpick halves into the head and center of the stem. Push the toothpicks deeply into the flower so they will be covered by the petals.
- Arrange miniature flowers in a shallow vase or a bowl of well-watered sand.
- Add height to an arrangement by placing filler in the bottom of the vase (bubble wrap, crumpled chicken wire, florist's foam, marbles, pebbles, Styrofoam pellets, or colored fish-tank gravel).
- Add fillers to give a bouquet an airy look. Use fern asparagus, ivy, Queen Anne's lace, statice, or any foliage that's available in the garden.

- Save the petals that drop from aging flower arrangements. Lay the petals face down on a paper towel, let them air-dry for two to four days, and then display them in a low glass bowl as potpourri.
- Tuck clippings of scented geranium leaves, especially rose-scented kinds, in among the flowers of a non-fragrant flower arrangement.

***FLOWERS, CUT, longest lasting***

- The longest-lasting cut flowers are aster, Canterbury bells, chrysanthemum, clarkia, cornflower, delphinium, freesia, gerbera, gladiolus, hydrangea, lily, and orchid.

***FLOWERS, CUT, most fragrant***

- The most fragrant cut flowers are carnation, freesia, hyacinth, lily, lily of the valley, nicotiana, peony, phlox, rose, stock, and sweet pea.

***FLOWERS, CUT, to prolong the life of***

- Pick flowers early in the morning, and don't take any more leaves than necessary. After picking, use a sharp knife to cut a two-inch slit up the stem, and then cut across the bottom at a forty-five-degree angle. Try using a sharp pocketknife to cut the flowers; sometimes it's easier to use than a regular knife.
- Scrape the bark on woody plants from the last two inches of the stem base until green shows; cut across the bottom of the plant at an angle, and then make a two-inch slit up the stem. Another method is to pinch off the bottom of the stem base with wire snips, which pulverizes the end.
- Cut flowers when they're half open, submerge the stems and flowers in warm water (about 100°F) for thirty minutes; then arrange them in a vase of cool water, and let them air-dry. Warm water contains fewer air bubbles and is more readily absorbed. (Air bubbles can get trapped in the stems and block the uptake of water to the upper foliage and flowers.) Warm water also works best to dissolve floral preservatives.

- Avoid using water that's been treated by a water softener. Draw the water from an outside faucet.
- Refrigerate cut flowers at night to nearly double their life. Make sure there is no fruit in the refrigerator when you do this. Fresh fruit gives off ethylene gas, which is detrimental to cut flowers. Don't put roses in the refrigerator; it can get too cold for them.
- Keep the water in the vase fresh, and have the flowers last longer by adding any of the following to one quart of water: a pinch of baking soda, a crushed aspirin, one tablespoon of activated charcoal, one-half teaspoon of medicinal-type mouthwash, one-quarter teaspoon of chlorine bleach (do not use bleach if the flower vase or container is silver or pewter).
- Lengthen flower stems that have been cut too short by inserting each stem into a flexible straw cut to the desired length.

### *FLOWERS, CUT, WILTED, to restore*

- Revive cut flowers by recutting their stems in warm water.
- Perk up cut flowers with a bath. Lay stems, leaves, and flowers in warm water for half an hour before arranging them.
- Add one teaspoon of mild liquid dish soap to the water in a vase of wilted flowers.

### *FOXGLOVES, hints on*

- Plant foxgloves under the shade of trees. These self-sowers will quickly establish themselves.
- Cut off the main spike when the blossoms fade. New shoots will develop for another round of growth.

### *FREESIAS, hints on*

- Choose single-flowered kinds of freesias for the most fragrance. They are more fragrant than the doubles. And the white-flowered kinds are the most fragrant of all.

- Bury the corms so they are covered by two and a half to three inches of soil, and space them three to four inches apart in full sun.
- Plant freesias with annuals or perennials that have strong foliage for the plants to lean on. They will help to prop up the freesias.
- Give freesias little or no fertilizer, and water established plants deeply but infrequently.

### *FROST CONDITIONS, to protect plants from*

- Lay newspaper over low-growing plants, prop paper bags over small plants (anchor them with stones), and cover larger plants with buckets, baskets, plastic containers, cardboard boxes, or even Styrofoam coolers.
- Cover large planted areas with bed sheets, tablecloths, blankets, or beach towels.
- Put double-row covers or plastic tunnels over the rows. (Make sure there is a thin layer of air between the plants and the plastic tunnels.)
- Protect plants from light frost by wetting them with a fine spray from the garden hose before you go to bed.
- Hose off any frost on the plants early in the morning, before the sun hits their leaves. This will minimize the damage.
- Put container plants in the garage or under the shelter of a front-porch roof, eaves, or a south or west wall with an overhang.

### *FUCHSIAS, hints on*

- Prune fuchsias in early spring before the plants begin to grow.
- Pinch off fuchsias' soft growing branch tips, just above a set of leaves, to force growth into side branches and to prevent them from becoming leggy.

### *FUNGICIDE SPRAY, to make*

- Mix one tablespoon of baking soda, one teaspoon of mild liquid dish soap, and one gallon of water. Spray the plant, paying special attention to the undersides of the leaves. Do this in the morning so the foliage has a chance to dry. Use the spray for powdery mildew, leaf blight, leaf spot, gummy stem, anthracnose, fusarium and verticillium wilts, and early blight on tomatoes and potatoes. Apply the spray every five to seven days. The USDA has approved baking soda as a fungicide.

- Mix one-quarter cup of low-fat milk or whey, four drops of mild liquid dish soap, and one gallon of water. This is good for squash and cucumbers. Apply the spray every ten days, using a freshly made batch each time.

### *FUNNEL SUBSTITUTE*

- Cut the bottom off a plastic bottle or the corner off a plastic bag to make a funnel. Different sizes of bottles and bags will make different-size funnels.
- Cut the corner off an envelope at an angle. Use as a small funnel for seeds.

# G

### *GARDEN EQUIPMENT, hints on*

- Check whether a garden center or nursery will lend you the equipment you need free of charge. For example, if you are planting a lawn, ask whether you can borrow a seed spreader when you buy your seed; if you are installing sprinklers, ask if you can borrow the necessary tools.
- Check garage sales, thrift stores, classified ads, or Craigslist for free or inexpensive equipment.

### *GARDEN HOSE, hints on*

- Coil the garden hose while the water is still running through it so it doesn't kink up.
- Use a hose hanger, which will prevent the hose from crimping and cracking.
- Prevent the hose from knocking down, crushing, or damaging plants while in use by inserting short lengths of PVC pipe at the corners of the garden beds.

### *GARDEN HOSE, to repair*

- Sand the surface of a rubber patch and the area around the hole very lightly with fine sandpaper. Apply contact cement to both surfaces, and wait until the cement dries. Then firmly press the patch over the hole, and wrap the patched area with either silicone tape or electrical tape.
- Treat a small leak by wrapping the cleaned and dried area with silicone tape or electrical tape, overlapping it a few times.

**GARDEN HOSE, OLD, *to repurpose***

- Turn an old hose into a soil-soaker hose. Punch holes at regular intervals along the length of the hose on one side using a heated ice pick, a sharp knife, or a screwdriver. Tie up the end, or just leave a closed shut-off nozzle on the end. You can shorten the hose by closing off the end with a clothespin or a C-clamp. When using the hose, lay it on the ground with the holes facing down.
- Use a piece of the hose to wrap around and anchor a newly planted sapling.

**GARDEN MISCELLANY, *hints on acquiring***

- Check out salvage yards for paving stones, gates, fences, and masonry ornaments.
- Use broken chunks of concrete paving as a natural-looking substitute for rocks or landscaping blocks.
- Keep an eye open for houses and gardens being demolished by wreckers. Ask if you can cart away the bricks or any other items you can use in your garden.

**GARDENIAS, *hints on***

- Expose the gardenias to a minimum of five hours of sunlight a day.
- Treat leaves that are yellow between the veins with ferrous sulfate or iron chelate. Mix either ingredient into the top inch of soil in spring and summer. Make sure the soil is warm so the gardenias can absorb the treatment. Gardenias need heat, especially at their feet. Alternatively, treat the gardenias with a foliar application of iron.

**GARDENING CLOTHES, *hints on***

- Buy a pair of army fatigue pants from an army-surplus store. They are roomy and have drawstrings at the bottom, a reinforced seat and knees, and lots of big outside pockets to hold seeds and tools.

- Pick up a cloth nail apron from the hardware store, and wear it when planting. The many pockets will hold seed packages and tools.
- Wear subdued colors. Research has proven that bright yellow attracts aphids, bright blue attracts thrips and leaf miners, and red and orange attract leafhoppers, flea beetles, and other insects. Wearing these colors might well spread insects around the garden.
- Put a drop of lemongrass oil on the brim of your gardening hat to keep insects from bothering you while gardening.

***GARDENING GLOVES, hints on***

- When the gloves get holes, change them onto opposite hands, and wear the gloves backward.
- Purchase knit military gloves from an army-surplus store. These gloves are reinforced with leather or plastic, hold up well, and sell for just a few dollars.
- Store gloves in a self-sealing plastic bag so insects can't crawl inside.
- Use leather or gauntlet gloves or long oven mitts for cutting roses and for pruning roses and berry bushes. They offer more protection than regular gloves.

***GARDENING GLOVES FOR COLD OR WET WEATHER, hints on***

- Put a pair of extra-large rubber gloves over your garden gloves to keep your hands dry and warm.
- Check the army-surplus store for neoprene gloves, which are similar to wetsuit gloves used by divers. These gloves insulate your hands and keep them dry and warm.

***GARLIC, hints on***

- Plant garlic in the fall; garlic blooms in late spring to summer.
- Buy a few plump bulbs from the supermarket or farmers' market, and break them into individual cloves. Plant cloves with the pointed ends up about three inches deep and four to six inches apart.

- Stop watering about two weeks before harvesting, or when 25 percent of the leaves have turned brown. Harvest when 60 to 75 percent of the tops are brown, or when the tops yellow and topple over; then set the garlic out to dry in a shady place for about a week or so. Heads are cured when the skins and stalks feel completely dry.

***GARLIC-OIL SPRAY, to make***

- Soak three ounces (about six to eight cloves or one-third cup) finely minced garlic in one-quarter cup of mineral oil for one or two days. Add two cups of water and one teaspoon of insecticidal soap or mild liquid dish soap, such as Ivory or Dr. Bronner's. Stir the solution thoroughly and strain into a glass jar. Use two to four tablespoons of the solution to one quart of water, and spray it on infested plants. This formula is effective on aphids, cabbage loopers, earwigs, June beetles, leafhoppers, squash bugs, and whiteflies. Some studies also suggest that it has fungicidal properties.

***GERANIUMS, hints on***

- Pinch back new growth for bushier geranium plants with more blooms. Continue pinching bloomed-out branches throughout the summer to keep plants looking neat and to encourage new budding. Root the cuttings for new plants.
- Save the leaves of scented geraniums that you prune off or that fall off. Dry them on paper towels, and then use them for scenting drawers and closets or to make sachets. Or just put the leaves in a shallow bowl, and use as potpourri. Tuck cuttings, especially rose-scented kinds, among the flowers of an unscented bouquet to add fragrance.

***GLADIOLUS, hints on***

- Plant gladiolus in a circle so the plants will support each other and not require staking. They also make a more dramatic showing when planted this way.

- Fertilize the plants with cottonseed meal rather than wood ashes or bone meal if the soil is alkaline.

***GOPHERS, to control***

- Lay gopher wire or half-inch hardware cloth under beds to prevent gopher damage.
- Try using a garden hose to flood a gopher tunnel with water.
- Purchase gopher traps as the most effective remedy.

***GRAPES, hints on***

- Protect ripening grapes by covering each cluster with a paper bag stapled shut or secured with paper clips. Cut a small hole in the bottom of the bag to let out rainwater. The grapes will continue to ripen in the bag.
- Harvest the leaves from the upper third of the vine in June, when the leaves are young and tender. Or use less tender leaves to make dolmas. Use immediately or freeze for later use.

***GRASS, hints on***

- Buy grass seed recommended for your area. Your local extension service will give you advice on the best blends. Or request information from the Lawn Institute at thelawninstitute.org.
- Pretreat the seed to speed germination. Mix two tablespoons of strong cold tea into each pound of seed, cover the seed, and leave it in the refrigerator for five days. Dry the seed for one or two days on newspapers laid on the garage or basement floor, and then sow the seed as usual.
- Seed in the fall months for best results. Temperatures are cooler, and most annual weeds have finished their growth cycles.
- Use the insert from an old salad spinner to help scatter the grass seed.
- Cover seeded areas with burlap, old flannel sheets, muslin, or any fabric that will absorb moisture. Keep the covering moist until most of the seeds have germinated, then carefully remove it without disturbing the

roots. Covering the seeded area keeps the soil moist and prevents birds from eating the seed.

- Mow the grass in the morning before the sun is up, and cut off no more than a third of the blade. Let the grass grow two or three inches before cutting. (Longer grass prevents sunlight from reaching weed seeds, and the grass is better able to resist wind, sun, and drought.) Leave the clippings on the lawn to provide nitrogen.
- Prevent grass from sticking to the blades of a hand mower by spraying the blades with nonstick cooking spray or rubbing them with oil.
- Keep a sharpened spare blade in the garage or tool shed. When the blade on the mower gets dull, swap it for the sharp blade, and take the dull one to be sharpened. This way your mower is always ready for service.

***GRASS AND WEEDS IN PAVING, PATIO CRACKS, AND WALKWAYS, to remove***

- Pour boiling water from a teakettle onto the grass and weeds.
- Pour white vinegar or table salt in the cracks.

***GRASS AND WEEDS IN UNPLANTED SOIL, to remove***

- Water the unplanted area well, and then lay plastic over the top. Leave it in place until the grass and weeds underneath turn yellow and die, usually within a couple of weeks.
- Put cardboard or layers of newspaper over the area. Wet the cardboard or newspaper and spread grass clippings or mulch over the top. When the weeds or grass decompose, either remove the cardboard or papers or leave them to decompose.

***GRASS SEED, to test for germination***

- Sprinkle a teaspoon of seeds on top of a small container of weak room-temperature tea, and place it on a sunny windowsill. If the seed is viable, it should germinate in about a week. If nothing happens, the seed is too old and should be discarded.

- Plant some seeds in potting soil; water and see how well it germinates; or place the seed between damp paper towels, seal in a plastic bag, and see how well the seed geminates. If the seed germinate rate is less than normal, sow the seeds at a thicker rate. If nothing happens after one week, discard the seed.

### *GRASSHOPPERS, to control*

- Handpick grasshoppers in the early morning during spring and summer.
- Observe their evening behavior to locate their roosting sites, usually in hedges and shrubs, then spray in those areas after dark with a garlic-oil spray, or pick grasshoppers off the plants. They are easy to catch at night, as they don't hop.
- Fill one- or two-quart jars with a solution of one part molasses mixed with ten parts water, and then place the jars where the infestation seems worst. Empty and reuse the traps.

### *GROUND COVERS, best for shady areas*

The best ground covers for shady areas are bugleweed, bunchberry, creeping phlox, foamflower, ivy, lady's-mantle, lilyturf, sweet woodruff, wild ginger, and winter creeper (winter creeper can be invasive).

### *GROUND COVERS, hints on*

- Prevent invasive ground covers such as English ivy from encroaching into lawns or flower beds by installing steel or plastic edging buried six to eight inches deep.
- Start new plants of ivy, myrtle, and others by pinning the vines to the ground with bobby pins.
- Pick plants with the longest shoots, and then take a runner, and cover a leaf axis with a little soil. It will root and start a new plant. Weigh the soil with a stone if necessary.

# H

### *HAND CARE WHEN GARDENING, hints on*

- Rub a thin coat of petroleum jelly on your hands before starting a messy job.
- Scrape your fingernails over a bar of soap before you begin gardening. This makes nail cleaning easier.

### *HANDS STAINED FROM GARDENING, to clean*

- Put a generous amount of hand lotion or cooking oil in your palms, add a little sugar, rub hands well, and then wash with soap and warm water. A faster way to clean lightly stained hands is simply to add a little sugar to the soapy lather when you wash them.
- Moisten a couple of tablespoons of dry oatmeal or cornmeal with milk, vigorously massage it into the hands, and then rinse. This should leave your hands smooth and remove stains.
- Wet your hands, and scrub them with a dampened nail brush sprinkled with baking soda.

### *HEDGES, to trim in a straight line*

- Tie a string to a branch at one end of the hedge, and run it across to the other end, making sure the string is straight. Or stick two stakes in the soil at either end of the hedge, and tie the string from one to the other.

### *HERBS, hints on*

- Turn an old stepladder into an herb garden. Lay it flat in a permanent area, fill it with soil, and grow a different herb in each section.

- Plant herbs throughout the garden. They exist happily in flower and vegetables beds, and they some repel specific plant pests. Or grow herbs in nursery pots, and then move them around the garden as they are needed as pest repellents.
- Grow flowering herbs to provide food for beneficial insects. Let one or two of their favorites—anise, caraway, chervil, dill, fennel, or parsley—bloom at the end of the season.
- Grow herbs as fillers for cut bouquets. Yarrow, oregano, and rosemary blend in well with fresh and dried arrangements as well as provide food for beneficial insects.
- Avoid fertilizing herbs grown in the garden; fertilizer- or nitrogen-rich soil produces rampant growth at the expense of flavor.
- Grow tall herbs such as coriander, dill, and tarragon against a wall or fence or toward the back of the flower bed.
- Place low-growing herbs such as marjoram, parsley, and thyme where they will not be overshadowed by other plants. Or plant these herbs in pots, hanging baskets, or window boxes, or as edgings for flower beds.
- Choose French tarragon for cooking; it's sold as transplants. Russian tarragon, which is grown from seed, has little flavor.
- Try planting Greek oregano for cooking. It is stronger than most oreganos. For milder flavor, use Italian oregano.
- Cut the bottom out of a plastic flowerpot or plastic kitchen container, and invert it over an herb that tends to get floppy, such as parsley, chives, or oregano. The collar will support the stem, prevent the leaves from touching the soil, and keep the plant clean.
- Keep foliage fresh and at peak quality for cooking by cutting the herbs back regularly; however, don't cut perennial herbs back more than one-third at any time.
- Harvest herbs with scissors to avoid crushing them with a dull blade, which can alter their flavor. Or simply pinch the leaves off of the plant.

***HOLLYHOCKS, hints on***

- Change the hollyhocks' planting location yearly to avoid rust (a common disease of hollyhocks), and pull out plants that have bloomed their second year to reduce the prospect of rust; young plants are less susceptible.
- Control rust by
  1. avoiding overhead watering;
  2. dusting the plants early in the season with sulfur; and
  3. removing the first leaves on which rust appears. (Do not apply the sulfur during hot weather over 80°F.)
- Cut off the flower stems just above the ground after the blooms fade. Continue to water and fertilize the plants, and they will produce another crop of flowers. Make sure you fertilize them a few times during the regular growing season so they'll have the strength for a second harvest. Conversely, don't over fertilize, which may lead to soft stems.
- Chose the dwarf variety of hollyhocks for small gardens or patio containers. They have double blooms, which more than compensates for their lower height.
- Cut hollyhocks in early morning or late evening, when the flowers are not yet fully open. Singe the cut ends with a match until they blacken, or dip the ends in boiling water for several minutes. Repeat every time you recut the stems. This breaks the milky seal the flowers secrete.

***HORSERADISH, hints on***

- Confine horseradish to its own bed; it is invasive. Another method of planting is to confine it in a five-gallon pail with the bottom removed that you have sunk into the soil (see *BUCKETS, hints on*).
- Use the tender inner leaves of horseradish in a tossed salad.
- Harvest pieces of horseradish root from the outside of the root clump as needed, any time through fall, winter, and spring. That way you'll have horseradish that is fresh and hot.

- Freeze the horseradish for convenience and ready availability. Grind the peeled root in a food processor, and then puree it with a little water and vinegar. Freeze the puree in ice-cube trays, and then store flat in a freezer bag after pressing out the air.
- Propagate horseradish by root cuttings in spring or fall.

***HORTICULTURAL OIL SPRAY, to make***

- To make horticultural oil spray, add one tablespoon of mild liquid dish soap, such as Ivory or Dr. Bronner's, to one cup of light vegetable oil, and shake it vigorously. Add one to two teaspoons of the mixture to each cup of water, and shake again to emulsify it. Pour mixture into a spray or pump bottle, and apply at ten-day intervals. This formula is effective against many plant pests, including spider mites, armored and soft scale, mealybugs, psyllids, whiteflies, aphids, leaf rollers, web-worms, cankerworms, corn earworms, and leafhoppers (see *OIL SPRAY, hints on* for precautions).

***HOT CAPS (cloches to protect tender plants during cold weather), hints on***

- Use clear-plastic gallon milk bottles or two-liter soda or water bottles. Remove the labels, and cut the bottoms off. Place one over each plant needing protection, and leave on at night. During the day, remove either the bottle cap or the entire bottle, depending on weather conditions. Put stakes through the tops, if necessary, to prevent the bottles from blowing away.
- Collect suitable-size boxes, and use them as hot caps.
- Inquire at restaurants, delicatessens, or local school cafeterias for their leftover half-gallon or gallon glass or plastic jars that pickles, olives, and other products are packed in. Prop up one edge of the jar during the day, or simply remove them now and then.

- Bend metal coat hangers or wire into hoops, and position them along the row or seedbed needing protection, sinking them slightly into the ground. Cover the hoops with plastic that is wider than the wire, and anchor the edges with stones or soil.

***HUMMINGBIRDS, to attract to the garden***

- Attract these birds to the garden by planting deep tubular flowers in red, orange, or fuchsia colors. The birds are particularly fond of honeysuckle, red-hot poker, bee balm, and cypress vine. Hummingbirds consume more than half their weight in nectar and insects every day and can help keep garden pests under control.

# I

## *INFORMATION ON GARDENING, to obtain*

- Call your local county extension or cooperative agent for information about gardening. This branch of the Department of Agriculture offers advice, booklets, and information over the phone. Master gardeners can answer questions on gardening from the general public. Call for free advice. (Look in your local phone book for the Agricultural Department, listed in the "County Government Offices" section.)
- Visit botanical gardens, especially when they have exhibits and plant sales, and ask questions. Or attend lectures by gardening experts put on by local nurseries, and ask questions.
- Use your local library for gardening books, magazines, and research information.
- Build a reference file from information garnered from seed catalogs, the gardening section of the newspaper, and other various sources. Arrange them alphabetically for easy reference.

## *INSECTICIDAL POWDER, to apply*

- Apply insecticidal powder early in the morning, when there is little or no wind and when dew is present to hold the dust on the foliage.
- Dust the plants thoroughly so the powder covers every area of the plant, including the undersides of the leaves.
- Cover small plants with two or three sheets of newspaper, and spray up underneath. This will control the powder. Cover shrubs and fruit trees with pieces of plastic.

**INSECTICIDAL SOAP, *information on***

- Make insecticidal soap by mixing one to two tablespoons of mild, uncolored liquid dish soap, such as Ivory or Dr. Bronner's Castile, with one gallon of water.
- Mix commercial insecticidal soap with distilled water if the water in your area is hard. The soap is less effective in hard water and can also cause leaf damage.
- Use insecticidal-soap spray to control aphids, cucumber beetles, earwigs, gnats, grasshoppers, leaf hoppers, mealybugs, psyllas, rose slugs, scale insects, spider mites, spittlebugs, squash borers, thrips, and whiteflies.
- Avoid using insecticidal soap on ferns, palms, gardenias, and nasturtiums, which are easily damaged by soap sprays.
- Avoid using insecticidal soap on plants under water stress.
- Spot-spray a small area of the plant before applying insecticidal soap or any pest-control spray. Then wait a day, and check for any damage or tip burns. Spray early in the morning, before the bees become active and before the sun hits the plants.
- Hit the pests directly with the spray. Limit the spray to a small area where you can actually see the pests to avoid harming beneficial insects.
- Wash the soap off after ten to thirty minutes to prevent the buildup of fatty acids from the soap.

**INSECTICIDAL SPRAYS** see *INSECTICIDAL SOAP; GARLIC-OIL SPRAY; HORTICULTURAL OIL SPRAY; PEPPER SPRAY*

**INSECTS, BENEFICIAL, *hints on***

- Create a hospitable garden environment for beneficial insects by eliminating the use of pesticides. If you do use preventive measures, do so only on the most ailing, pest-infested plants and only during early morning or evening, when bees are not active. Avoid spraying plants in bloom.
- Plant a diversity of flowers to provide a variety of food sources. Honeybees forage on a long list of common plants. Pollen- and nectar-producing

flowers are a food source for beneficial insects, and often the adult insects will feed only on nectar or pollen, while their young will eat the pests.

- Allow an herb or two—such as anise, caraway, chervil, cilantro, dill, fennel, or parsley—to bloom, or let an old carrot, celery, or parsnip plant go to flower; any of these plants will encourage beneficial insects to make themselves at home in the garden.
- Have some bushes, shrubs, or hedges to provide shelter and protection for lacewings and other beneficial insects.
- Encourage soil-dwelling beneficial insects by adding compost to the soil.
- Provide hiding spots for soil-surface beneficial insects by leaving some stones and undisturbed mulch in place.
- Drill a few holes in a piece of soft wood, and hang it in a tree to encourage bumblebees to nest on the property. Cosmos also attracts bumblebees and other solitary bees.
- Give a few desirable weeds a place in the garden: common sorrel, goldenrod, tansy, Queen Anne's lace, or wild daisy. These weeds attract beneficial insects.
- Sow some native wildflowers. Or try to include some single-flowered plants in the garden, such as Arabis (a.k.a. rockcress), Echinacea (a.k.a.coneflower), Scabiosa (a.k.a. pincushion flower), or Veronica (a.k.a. speedwell). Particularly attractive to beneficial insects are daisy-like flowers or plants with clusters of small flowers, such as dill and fennel and many common herbs.
- Plant any of the following to attract bumblebees, butterflies, honeybees, lacewings, ladybugs, and other airborne beneficial insects:
    1. Herbs: angelica, anise, anise hyssop, bee balm, borage, caraway, chervil, cilantro (a.k.a. coriander), comfrey, dill, fennel, lavender, mint, oregano, sage, rosemary, and thyme.
    2. Flowers: aster, baby blue eyes, baby's breath, black-eyed Susan, blanket flower, butterfly bush, candytuft, coreopsis, cosmos, foxglove, goldenrod, hollyhock, marigold, nasturtium, single dahlia, statice,

sunflower, sweet alyssum, tidy tips, verbena, white evening primrose, wildflowers, and yarrow.

***INSECTS, BENEFICIAL, PURCHASED, hints on***

- Use a magnifying glass to get a good look at the insects you're going to purchase so you'll be able to recognize them in the garden. To see photos of the insects, Google "beneficial insects."
- Avoid spraying or using sticky insect traps once you have released the insects into the garden.

***IRON FOR PLANTS, hints on***

- Spray the iron-deficient foliage with iron sulfate, or add chelated iron to the soil. Over fertilizing with phosphorus, which ties up the iron, can also create an iron deficiency in plants.
- Make an iron tonic for the plants by using old nails and tools, iron scraps, and anything rusted beyond repair. Let the items sit in a bucket of water until they are thoroughly rusted, and then use the rust water for the plants. Add one-quarter to one-half cup of the rust solution to each gallon of fresh water, and pour it around the bases of the plants.

# J

***JAPANESE BEETLES, to control***

- Go looking for Japanese beetles early in the morning, before they are ready to fly. Knock or shake them from the plants into a container of soapy water.
- Plant white geraniums among your favorite plants. The beetles are attracted to the geraniums and are then easy to collect and destroy. Four-o'clocks also have the same effect on the beetles.

# K

### *KALE, hints on*

- Plant kale in flower beds or a container. Kale makes a decorative plant.
- Use the tender young leaves for salad and the older leaves for stir-frying or steaming. Harvest the outer leaves as needed. When growth slows down and the outer leaves are tough, use the inner tender core for salad.
- Leave this hardy winter vegetable in the ground through the winter; it's best touched with a bit of frost, and you can harvest it from under a cover of snow.
- See also *CABBAGE PESTS, to control.* Kale shares the same pests and diseases.

### *KNEELING PAD, to make*

- Wrap a piece of foam rubber, an old cushion, a pillow, or a three-inch stack of newspaper in a plastic bag or old pillowcase to make a kneeling pad. Or make a cover from an old bath towel if you garden in shorts. Wash the cover as needed.
- Sew large patch pockets into the knees of your gardening pants, then slip a piece of foam into each pocket. Remove the foam when laundering the pants.
- Look for slip-on knee pads at a sporting-goods store.

### *KOHLRABI, hints on*

- Use kohlrabi as a salad green. Kohlrabi is more tolerant of heat than other cole crops and can also thrive in semi shade and cool weather.
- Plant kohlrabi with onions and beets but not with tomatoes, peppers, or pole beans, which are not beneficial companions.

# L

**_LAVENDER, hints on_**

- Grow lavender where there is good drainage, and don't overwater or over fertilize.
- Cut the flowering stalks periodically for flower arrangements, or shear them regularly to keep them tidy. This will also promote new blooms.
- Cut back plants that grow larger than three feet by as much as half; this will prevent them from turning woody. Start this no later than the second year in early fall.
- Plant English lavender for fragrance and drying.
- Dry the lavender by hanging the spikes in bunches upside down in a paper bag. The flowers will drop into the bag when they are dry. (Cut the stems in dry weather when about half the flowers on the spike are open.)

**_LEAFHOPPERS, to control_**

- Set out yellow sticky traps to control leafhoppers. Or use empty bright-yellow containers: coat them with motor oil, petroleum jelly, or Tanglefoot, and set them near infested plants. Remove the insects every week or so, and renew the sticky coating.
- Spray the plants with insecticidal soap in early morning or evening. Spray directly on the leafhoppers; insecticidal soap works on contact and has no residual effects (see *INSECTICIDAL SOAP*).

**_LEAVES, COLLECTING, hints on_**

- Lay an old plastic shower-curtain liner, a plastic tablecloth, or a sheet on the ground, and rake leaves onto it; then gather up the four corners, and drag it to the compost pile. Use a large opened-up plastic garbage bag for smaller amounts.

- Clip a garbage bag to a fence with clothespins, then hold it open with one hand, and shovel in the leaves with the other.
- Fit a large bag into a garbage can, and lay the can on its side. Shovel in the leaves until the can is about one-quarter full, then remove the bag, and set it upright. (The leaves will form a base so the bag sits solidly on the ground.) Then continue filling the bag with leaves.

***LEEKS, hints on***

- Avoid planting leeks with beans or peas; leeks are not beneficial companions for legumes.
- Plant leeks with carrots to help repel carrot flies or with roses and other susceptible plants to repel aphids
- Set transplants in the bottom of a four- to six-inch narrow trench. As the plants grow, gradually fill the trench with soil or mulch. In this way, four inches or so of the stem beneath the covering will be blanched. Otherwise, blanch the leeks when they're almost fully grown by tying newspaper or empty paper towel rolls around them.

***LETTUCE, hints on***

- Refrigerate lettuce seeds overnight for quicker germination in warm weather.
- Sow romaine seeds a little thicker than normal; they don't germinate as well as other types of lettuce.
- Promote quick, succulent growth by providing adequate moisture and fertilizing once or twice during the growing season. Another fertilizer option is to foliar feed with seaweed extract diluted to half strength every two weeks. Slow growth makes lettuce tough and bitter.
- Squeeze lettuce into empty spaces around the garden or use it to edge a flower bed.
- Protect lettuce with row covers if flea beetles or leaf miners are a problem. Apply the protection right after seeding (see *ROW COVERS, hints on*).

- Shade the plants as needed with row covers that are propped up at the sides, a shade cloth, or a piece of old sheeting suspended on tall stakes. Or plant the lettuce in the shade of taller crops, such as under a pole-bean trellis or tepee.
- Spray lettuce leaves with a fine mist of water at least once a day if the weather is hot or windy.
- Pick the outer leaves of loose-leaf lettuce when they are about three to four inches long. You can safely pick down to about the six center most leaves.
- Cut lettuce plants off about one inch above ground level when they are three to six inches tall. The lettuce will produce another harvest in two to three weeks.
- Give head lettuce a dose of nitrogen and lots of water close to harvest time. Two-thirds of the head's weight develops during the last one-third of its growth.
- Harvest lettuce in the morning, when it's at its crispest state.

***LILIES, hints on***

- Cut lilies when their buds show color or when they're partially open, and cut only the short stem that attaches the flower to the main stalk. Split the stem ends, and then place them in lukewarm water.
- Prevent saffron-colored pollen stains from getting on cut flowers by removing the anthers before arranging.
- Do away with pollen stains completely by planting double-flowered varieties, which have no pollen. Their anthers and stems have been converted to petals. Another alternative is to plant a sterile single-flowered variety, such as hodge podge.

***LUPINE, hints on***

- Soak lupine seeds in warm water overnight, or nick the seeds with a file to hasten germination.

- Sow the seeds where they are to grow in clumps of three to four seeds that are twelve to sixteen inches apart.
- Clear away any mulch when the flowers are going to seed. Lupines are self-sowers.

# M

***MELONS, hints on***

- Plant melons where the vines will get reflected heat from a wall or fence; melons are heat lovers.
- Look for bush-type melons if you need to conserve space; melons need lots of space.
- Water to a depth of twelve inches, and keep a constant supply of moisture at the root zone. Before planting, sink a clay pot or a large coffee or juice can with holes poked in the bottom in the melon hill. Fill the container with water every day. Another watering option is to sink PVC pipes in the soil and fill them with water as needed.
- Fertilize early in the season, but not when fruit begins to develop.
- Pick off the fuzzy growing ends of the melons when there are plenty of melons on the vines. This will help the plant concentrate its energy into producing fruit rather than leaves.
- Support trellised melons by putting them in slings made from mesh bags from onions or fruit. Use strips of soft cloth (old sheets or pillowcases) or nylon hose to tie the vines to the trellis. Rig the slings so that the weight of the fruit is off the stem. Apply the slings when the melons are about the size of softballs.
- Keep untrellised ripening melons off the ground to prevent rotting. Mulch them with clean straw or cardboard, or rest the melons on flat stones or empty upside-down cans pushed into the soil. This will also keep the melons warm, keep them out of the shade of the foliage, and hasten maturity. Wrapping them in newspaper or black plastic will also aid the ripening process. The black color will absorb heat during the day and release it during the night.

- Water less frequently as the fruit nears maturity, and then stop watering eight to ten days prior to harvest. This allows the fruit to develop sugar and sweetness. Almost half of a melon's final sugar content develops during the last week of maturation.
- Allow the melons to ripen for another day or so after picking to enhance the flavor.

***MESCLUN, hints on;*** see also *LETTUCE, hints on*

- Interplant mesclun with tall-growing crops for shade in early summer.
- Make your own mesclun mix by combining several different kinds of loose-leaf greens that you particularly enjoy, and then supplement them with the more common lettuce varieties. Mesclun seeds keep for several years.
- Plant every week, and cut the tender greens while they are young.
- Keep mesclun thinned so none of the leaves touch each other; crowding slows growth. Use the thinnings in salads.
- Foliar feed with liquid seaweed extract biweekly for fast, succulent growth.
- Treat mesclun like lettuce by harvesting the outer leaves rather than the whole plant.

***MESH BAGS (FRUIT OR ONION), uses for***

- Open up mesh bags, string them together, and suspend them over a strawberry patch to prevent birds from eating the strawberries.
- Store bulbs or onions in mesh bags during the winter. Hang them high in a cool, dry place.
- Support trellised melons or other heavy fruit using mesh bags. Use them as slings, and tie them to the support using pieces of soft cloth or nylon hose.

***MILDEW, POWDERY;*** see *DISEASES, PLANT, hints on*

**MINT, hints on**

- Keep a mint plant in the garden to attract honeybees, hover flies, parasitic wasps, and other beneficial insects. Grow mint plants in sunken pots to prevent spreading.
- Plant mint with cabbage, broccoli, Brussels sprouts, and kale to repel cabbage butterflies, with eggplants and other vegetables to repel flea beetles, and with tomatoes to repel tomato hornworms.
- Use long sprigs of mint in flower arrangements. They blend in particularly well with daisies.
- Cut off mint flowers when they form, and cut back the plant in summer and fall to keep it looking neat.
- Take stem cuttings whenever plants are growing vigorously, and divide plants every three to five years.

***MOLES, to get rid of;*** see also *GOPHERS, to control*

- Place mole traps only in active tunnels. Stamp down upraised mole tunnels. If they get raised again, it shows the tunnel is active.
- Soak the tunnels and entrances with a castor-oil solution. Mix one-quarter cup of castor oil and two tablespoons of mild liquid dish soap with one gallon of warm water. Moles do not like the taste, smell, or feel of castor oil.
- Soak paper towels or rags in peanut oil or bacon grease, and push them into the holes or tunnels. As they become rancid, they will act as a deterrent.
- Place pinwheels at intervals in the lawn to scare moles away.
- Flood mole tunnels. This works, but it may take a while.

***MOSS AND ALGAE ON WALKWAYS, to remove***

- Scrape moss and algae off walkways with an old putty knife or a spatula.
- Scrub the walkway with a solution of one part chlorine bleach to four parts water. Make sure the bleach does not come in contact with any vegetation. Hose the walkway, and sweep away the dead moss.

- Mix equal parts white vinegar and methylated spirit, and apply it to the area with a scrubbing brush. Leave it for fifteen minutes, and then scrub the surface again with the mixture. Leave it for another fifteen minutes; then hose the area, and sweep away the moss with a broom. Another option is to paint the moss with straight methylated spirit, let it sit for fifteen minutes, scrub away the moss with a broom, and hose off the area.

### *MUDDY CONDITIONS IN THE GARDEN, **hints on***

- Avoid working in the garden when the soil is wet. If you absolutely have to do something, such as harvesting, stay off the grass so as not to compact the soil, and lay a board over the area where you need to stand.
- Use a plastic kneeling pad (see *KNEELING PAD, to make*).

### *MULCH APPLICATION, **hints on***

- Weed thoroughly before applying mulch.
- Spread the mulch two inches deep to suppress weeds. A four-inch layer is even better.
- Avoid letting the mulch touch the stems and crowns of plants or the bark of trees. Keep mulch one or two inches away from crowns and stems and six to twelve inches away from trunks.
- Refrain from using nutrient-rich mulch around drought-tolerant or native plants. Ask for a nutrient analysis if you're buying mulch from a garden center.

### *MULCH MATERIAL, **hints on***

- Use newspapers as mulch. Lay them in vegetable or flowers beds, wet them, and cover them with rocks, grass clippings, or soil. Or place several thick overlapping layers of newspaper on garden paths to keep them weed-free. Wet the newspapers down, and cover them with shredded bark or pine straw. Newspaper is an inexpensive, water-conserving, biodegradable mulch that decomposes rapidly. (Avoid the glossy pages or color inserts; they may contain harmful inks.)

- Use large poultry bags, dog-food bags, or kitty-litter bags: Open a bag, cut down one side and across the bottom edge, and anchor it with dirt or rocks. Cut holes for transplants.
- Use paper-pulp egg cartons or cardboard packing boxes that you've broken down and flattened: lay them on garden paths, and cover them with shredded bark.
- Use a thin layer of grass clippings, no more than a half-inch deep, as mulch. Or dry them out on the driveway before applying them to the garden; then apply them two or more inches deep. For best results, mix the grass clippings with straw or leaves.
- Use salt hay as mulch if it's available. Regular hay may contain too many weed seeds, and in wet climates it attracts slugs.
- Use seaweed that has washed up on the beach as mulch. (Take only part of the plant, and leave the rest.) Rinse the seaweed to remove the salt.
- Use straw as mulch. A layer of straw will prevent soil from being washed onto the plants and protect them from soil-dwelling diseases. Light-colored mulches such as straw also reflect light and tend to keep soil temperatures cooler. Use straw around lettuce and other crops that benefit from cool soil. Obtain it free from a stable, if possible, or purchase certified weed-free straw.
- Use autumn leaves as mulch for flower beds and shrub borders. Run a lawn mower with a bagger over the leaves to chop them up. Do not use the leaves of acacia, walnut, California bay, juniper, camphor, cypress, or pittosporum for mulch. When decomposing, these leaves release chemicals that are toxic to plants.
- Use a living mulch. White clover grown between young squash or corn or under fruit trees attracts beneficial insects, adds nitrogen to the soil, and controls weeds.
- Use pea gravel or small rock chips as mulch for a drought garden. Avoid decomposed granite, which is a great incubator for weed seeds.
- Use sawdust or shredded bark as mulch, but don't add more than two inches a year. If nitrogen hasn't been added, add a high-nitrogen fertilizer

such as blood meal or cottonseed meal to offset the nitrogen that will be lost as the sawdust decomposes. Avoid cedar sawdust, which is toxic to some seedlings. Shredded bark is stringy and soft and packs down into a good water-conserving mulch. (Inquire at a tree-care company or a utility company whether wood chips or shredded bark are available free or for a nominal price.)

- Use flat stones as mulch around perennials. Arrange the stones around the plants, and lay the stones as close together as possible. Tests indicate that perennials that grow beside or among rocks are more productive, because their roots are cooler and the source of moisture is more consistent. Also use the stones as a permanent mulch around small trees.
- Use fully matured compost as mulch, except on most native plants (the nutrient content may be too high). Or call your county or municipal Department of Public Works to inquire whether free compost is available in your area and ask about its content. Keep an eye on the area mulched with the free compost to check whether any weeds sprout, and then take action immediately.
- Use spent hops as mulch. (Ask a local brewery for the hops.)
- Use fallen pine needles as mulch for acid-loving plants. As the pine needles decompose, they will very slightly acidify the soil and help build its structure.
- Use sheets of aluminum foil as mulch to double the yields of squash and corn. The reflected light under the plants keeps the soil as much as six to ten degrees cooler, repels aphids and thrips, and encourages plant growth.
- Use black plastic as mulch. The plastic absorbs heat, which warms the soil by five to six degrees and protects fruits such as cucumbers or strawberries from rotting. Cut holes in the plastic with scissors or a knife, and insert the transplants. As a precaution when using black plastic with melons and squash, apply a thin bed of straw between the fruit and the plastic to keep rot from developing.

- Use old carpet as mulch in the vegetable patch. Cut small holes to allow water to penetrate.
- Use red plastic or a coat of red paint on black plastic to boost the yield of tomatoes. Tests showed that tomatoes mulched with red plastic had a 20 percent higher yield than those mulched with ordinary black plastic.
- Use green plastic as mulch for early spring crops; it retains the day's heat longer.

***MULCH, WOOD-CHIP;*** see ***WOOD CHIPS, hint on***

# N

***NASTURTIUMS, hints on***

- Plant nasturtiums to repel flea beetles, whiteflies, squash bugs, cucumber beetles, and other pests around beans, cucumbers, melons, and squash, and to act as a trap for cabbage caterpillars and aphids. Give the flowers a head start before planting the vegetables.
- Grow nasturtiums to attract beneficial insects such as honeybees, hover flies, lacewings, ladybugs, and parasitic wasps.
- Add a spot of color in poor dry soil by sowing nasturtiums. The flowers will flourish and even provide fragrance to the area. Grow the climbing or trailing varieties to cover fences or unsightly areas.

***NEW ZEALAND SPINACH, hints on***

- Assist germination by either nicking each seed with a file; rubbing the seeds between sandpaper; or pouring boiling water over them and letting them soak for twenty-four hours.
- Cut three- to four-inch lengths from the tender growing tips and use only the young, succulent leaves.
- Keep the plants cut back constantly to prevent the leaves from becoming too old.

# O

***OIL SPRAY, hints on***

- Apply dormant oil spray before leaf growth begins. Read the instructions on the package regarding temperature conditions.
- Use ultrarefined oil spray if temperature fluctuations are a problem. It is easier to use and less temperature specific than dormant oil spray.
- Substitute canola oil for light horticultural oil in disease-control formulas. It works just as well and is recommended by the experts.
- Avoid using oil sprays within a month before or after spraying with sulfur or compounds containing sulfur.
- Avoid using oil sprays during peak flowering times when beneficial insects feed on nectar, and avoid spraying any blossoms.

***OKRA, hints on***

- Raise the percentage rate of germination by freezing okra's seeds. Drop one seed into each compartment of an ice-cube tray, fill it halfway with hot water, and freeze it for a couple of days before planting.
- Mulch the bed with straw four inches deep to conserve moisture and keep the pods clean.
- Pick the pods when they are two inches long. If you keep the pods picked regularly, the plants will keep producing from early spring to frost.
- Remove the bottom leaves after harvesting the first batch of pods. This will encourage the plant to produce more.
- Pinch off growing tips of plants when they are two feet tall to keep them short and bushy.
- Wear gardening gloves or slip plastic bags over your hands to keep them from itching when harvesting the okra.

- Keep okra fresh for twenty-four hours by moistening the pods and spreading them out in a cool place with good air circulation.
- Freeze whole raw pods in freezer bags without blanching. Cut the cap off each pod before cooking.

***ONIONS, hints on***

- Plant onions throughout the garden to reduce the chance of damage from onion maggots. This will also help other plants, especially beets and carrots, ward off pests. However, don't plant onions with peas or beans. They are not happy bedfellows; onions can retard the growth of legumes.
- Repel onion maggots by sprinkling sand or dry coffee grounds in the bottom of the row or planting hole.
- Trap onion maggots by planting old, sprouted, or soft onions between onion seeds in early spring. Pull out and destroy the onion traps two weeks after the bulbs sprout.
- Trap onion flies with a quart-size milk carton coated with petroleum jelly. Place it near the plants, and wipe insects off with vegetable or baby oil.
- Bend the onion tops over to a horizontal position or cut them off when they begin to dry and turn yellow. This will focus all the growing energy in the bulbs. Dig up the bulbs when all the tops are dead, and let the bulbs lie in the sun for a few days to dry and improve their storage quality.
- Eat the thick-necked onions first because they don't store as well.

# P

**PANSIES, *hints on***

- Shade pansy plants from the midday sun for optimum growth.
- Pick the blooms weekly to have the plants bloom for a longer period and prevent them from setting seed.
- Cut the pansies almost to the ground when they get leggy, and give them a dose of liquid fertilizer. This will promote new growth and further flowering.
- Propagate pansies by removing young side shoots and rooting the cuttings in coarse sand or vermiculite.

**PARSLEY, *hints on***

- Plant parsley with roses to protect against rose beetles or with tomatoes and asparagus, which will provide mutual benefit.
- Keep parsley plants pinched back to encourage bushier growth.
- Cut out the bottom of a plastic flowerpot or kitchen container, and invert it over the parsley. The collar supports the stem, prevents the leaves from touching the soil, and keeps the plant clean.
- Let a plant or two bloom to provide food for beneficial insects.

**PARSNIPS, *hints on;*** see also *CARROTS, hints on*

- Use fresh parsnip seeds that are not more than a year old. Seeds older than a year will not germinate.
- Soak the seeds for several hours to promote germination. Or sprinkle the seeds on prepared soil, pour boiling water over them, then cover the seeds with potting soil, and keep moist until germination occurs.
- Sow radish seeds in with the parsnips to act as a planting marker and to help aerate the soil when the radishes are harvested.

- Avoid over fertilizing, or parsnips will produce too many hair roots. Too much nitrogen also results in stringy, flavorless parsnips.
- Harvest the parsnips after a hard frost for optimum flavor. Leave them in the ground over the winter and pick as needed.

***PEAS, hints on***

- Avoid planting peas with garlic, leeks, shallots, chives, or onions, which have a negative impact on the growth of peas.
- Interplant peas with radishes, spinach, lettuce, or other early greens.
- Speed germination by soaking the seeds in room-temperature water for eight to twelve hours before planting.
- Inoculate seeds with nitrogen-fixing bacteria to increase the yield and help the plants create more nitrogen from the air. Check the inoculant for the expiration date, store in the refrigerator, and use as soon as possible. (Obtain the inoculant from a seed-supply catalog or garden-supply store.)
- Moisten the ground thoroughly before planting peas, and do not water again until seedlings have broken through the surface.
- Provide a trellis for the peas along the north side of the garden, and train the vines as they grow (see *TRELLISES, hints on* for other ideas). Or plant short bushy types of peas, and use twiggy branches as supports. Another option is to plant seeds thickly so the short vines will support each other.
- Pinch back the growing tips of the stems to thin out the vines a little and start them producing more abundantly.
- Pick the peas daily to encourage more growth, and pick them in the mornings to preserve their flavor.
- Freeze English peas by shelling them and popping them loose into a freezer bag. They will freeze individually since they are dry. Freeze snow peas by spreading the pods on a baking sheet, and then packaging them into freezer bags when frozen.
- Cut off the pea vines at ground level once they finish producing. The root nodules add nitrogen to the soil and aid in the growth of following crops.

***PEAT MOSS, hints on***

- Incorporate peat moss into the soil rather than spreading it on top as a mulch. Peat moss can act as a barrier and dry out the soil.
- Dampen the peat moss before adding it to the soil. Pour a little boiling water on the peat to get the process started, and then add cold water to dampen it thoroughly. Leave it for several hours before adding it to the soil. Another way to dampen the peat moss is to cut a small hole in the bag it comes in and let a hose drip in slowly for a day or two.

***PEAT MOSS SUBSTITUTE***

- Replace peat moss, which is a nonrenewable resource, with eco-friendly compressed coconut fiber (a.k.a. coconut coir or cocopeat). It can retain up to nine times its weight in water. Reconstitute it with warm water for quick results.

***PEAT POTS, TRANSPLANTING FROM, hints on***

- Remove the bottom of the peat pot (unless roots have already penetrated the pot; then leave it in place), slit the sides, and tear off the rim. Any protruding rim above the soil will act as a wick to draw water away from the plant. If there are not too many roots sticking through the bottom, remove the whole pot. This allows the plant maximum contact with the soil.

***PEPPER SPRAY (insect repellent), to make***

- Blend three or four cloves of garlic, one medium-size onion, one teaspoon of hot sauce, and one quart of water. Steep it overnight, and then strain. Dilute with three to four quarts of water before using.
- Combine one tablespoon of cayenne and six drops of mild dish soap such as Ivory or Dr. Bronner's in two quarts of water. Let the mixture stand overnight before using.

**PEPPERS, *hints on***

- Start transplants in peat pots or other containers (pepper roots don't like to be disturbed), and then bury them in the ground to the top pairs of leaves. Or trench them like tomato transplants (see *TOMATOES, hints on*).
- Plant peppers around the garden. They make attractive plants in the flower bed.
- Keep sweet and hot peppers some distance from each other, as they can cross-pollinate.
- Pinch off early blossoms so the plants can bush out before fruiting. This will increase the yield.
- Use peony rings as supports for large plants.
- Provide shade and humidity in hot climates by planting peppers close together, about eight to twelve inches apart. This will produce better yields and decrease the chances of blossom drop and leaf scald.
- Shelter the plants from intense sun by planting them alternately in a row with tomatoes. By the time the peppers are producing fruit, the tomatoes will be providing filtered shade.
- Give the pepper plants a dose of magnesium when they set fruit. Dissolve one tablespoon of Epsom salt in one cup of boiling water, and stir thoroughly. Add enough cool water to make one gallon, and pour it around the plants. Magnesium is a key nutrient for peppers.
- Spray light-green leaves with a solution of one teaspoon of Epsom salt dissolved in one pint of lukewarm water. The soil may be deficient in magnesium if adequate nitrogen is present.
- Keep peppers picked as soon as they mature to encourage more fruit. Cut them off, leaving a half-inch of stem on the pepper.
- If frost is predicted, pick all the fruit, or pull up the plants, and hang them indoors in a cool, dry spot until fruits ripen. Uprooted plants will also keep in a bucket of water in a cool location for a month.
- Dry hot chile peppers by pulling up the whole plant and hanging it upside down in a dry, airy spot. Remove the peppers when they are dry, and store them in a paper bag.

***PERENNIALS, hints on***

- Cut back perennial plants to two or three inches from the ground when the leaves start to look droopy or past their prime. The plants will leaf again. Do this before mid-July so the plants will have time to build root systems; in mild winter areas, you can wait until later.
- Shear back low-growing types after 80 percent of the blooms have faded. This will encourage new growth and initiate another bloom cycle. It will also do away with scraggly-looking plants in the flower bed.
- Use a hedge trimmer to deadhead plants with multiple flowers. The process will go much faster.
- Divide perennials every few years, or whenever they get overcrowded. Divide late or fall bloomers in spring, early or spring bloomers after they bloom or in the fall, and summer bloomers in early fall.
- Confine rapidly spreading and invasive plants in bottomless containers. Cut out the bottom of a five-gallon plastic bucket (well-washed joint-compound buckets, industrial paint buckets, and frosting or pie-filling pails from bakeries and doughnut shops make good candidates). Sink the container into the ground, leaving a one-inch rim above the surface, and keep runners trimmed so they don't invade the surrounding soil.
- Plant water-loving species with drought-tolerant ones with the help of polymer granules. When garden space is at a premium, you can plant both types by using the nontoxic granules, which absorb up to ninety times their weight in water and gradually release it.

***PERENNIALS, DROUGHT-TOLERANT***

- Drought-tolerant perennials include baby's breath, blanket flower, candytuft, chamomile, coreopsis, daylily, globe thistle, heliopsis, purple coneflower, spiderwort, Stokes' aster, stonecrop, yarrow, and yucca.
- Download a copy of "Nifty 50 Plants for Watersmart Landscapes," which is available through the San Diego County Water Authority. Go to watersmartsd.org, or simply Google the title.

### PERENNIALS, EASILY GROWN FROM SEED

- Perennials that are easy to grown from seed include beardtongue, black-eyed Susan, blanket flower, columbine, blue flax, blue salvia, lupine, pinks, primrose, purple coneflower, red hot poker, rose campion, Shaster daisy, and yarrow.

### PERENNIALS, FAST-GROWING

- Fast-growing perennials include alpine aster, baby's breath, bee balm, blackberry lily (a.k.a. leopard flower), blanket flower, cranesbill (a.k.a. hardy geranium), coreopsis (a.k.a. tickseed), creeping phlox, Cupid's dart, delphinium, dianthus, golden marguerite, horned violet, Japanese anemone, musk mallow, red valerian, rose campion, sea thrift (a.k.a. sea pink), Shasta daisy, snowdrop anemone, and veronica (a.k.a. speedwell).

### PERENNIALS, FAST-MULTIPLYING

- Perennials that multiply quickly include alstroemeria (a.k.a. Peruvian lily), Michaelmas daisy, oenothera (a.k.a. evening primrose), penstemon (a.k.a. beardtongue), physostegia (a.k.a. obedient plant or dragonhead), salvia, Shasta daisy, Stokes' aster, torenia (a.k.a. wishbone flower), and red valerian.

**PERENNIALS, GRAY WATER–TOLERANT;** see *PERENNIALS, TOLERANT TO SALTS AND ALKALINITY FOUND IN GRAY WATER*

### PERENNIALS, LONG-BLOOMING

- Long-blooming perennials include anise hyssop, autumn joy sedum, gold alyssum, black-eyed Susan, blanket flower, blue catmint, columbine, coreopsis, creeping phlox, hardy geranium, hollyhock, hummingbird mint, ice plant, lavender, Monch aster, perennial salvia, petunia, purple coneflower, purple rockcress, Russian sage, Shasta daisy, Stella d'Oro daylily, sweet William, tall garden phlox, veronica, and yarrow.

***PERENNIALS, LOW-MAINTENANCE***

- Low-maintenance perennials include agastache, black-eyed Susan, blanket flower, blue fescue, coral bells, coreopsis, hellebores, pasqueflower, penstemon (a.k.a. beardtongue), purple coneflower, salvia, sedum, and yarrow.

***PERENNIALS, QUICK-MATURING***

- Perennials that mature quickly include bee balm, coreopsis (a.k.a. tickseed), cranesbill, creeping phlox, gold-and-silver chrysanthemum, and Japanese anemone.

***PERENNIALS, SELF-SOWING***

- Self-sowing perennials include balloon flower, Siberian bugloss, columbine, common foxglove, four-o'clock, lobelia, purple coneflower, and red valerian. (At the season's end, remove the mulch from around the perennials so the seeds can reach the bare ground.)

***PERENNIALS, TOLERANT TO SALTS AND ALKALINITY FOUND IN GRAY WATER***

- Perennials that will be tolerant of the salts and alkalinity in gray water include African lily (a.k.a. lily of the Nile), California aster (a.k.a. silver carpet aster), coneflower, euryops (a.k.a. bush daisy), gazania, ground morning glory, gum plant, lamb's ears, lavender, lavender cotton, nasturtium, penstemon (a.k.a. beardtongue), pride of Madeira, sea thrift (a.k.a. sea pink), seaside daisy (a.k.a. beach aster), Saint John's wort, and yarrow.

***PERLITE SUBSTITUTE (lightweight inorganic soil amendment)***

- Use parboiled rice hulls, pumice, or broken-up Styrofoam to substitute for perlite. These items are sterile, lightweight, less expensive than perlite, and will not affect the pH of the soil.

***PESTS, PLANT, to control***

- Encourage beneficial insects to set up residence in your garden by providing a source of pollen and nectar for them (see *INSECTS, BENEFICIAL, hints on*).
- Attract birds to the garden by feeding them in winter and providing a birdbath. Birds are one of the best pest controllers and can kill thousands of insects in a day.
- Mix up plantings to confuse the bugs, or plant partial rows apart from each other.
- Know the enemy: If you see bugs you don't recognize, first identify them to determine the best method for handling them. For identification help, visit bugguide.net, sponsored by Iowa State University, or the Pest Management page on the North Carolina State Extension website: growingsmallfarms.org.
- Observe when insect infestations are worse on certain crops, and plant either a little earlier or later than you have in the past. This way, the plant's critical stage of growth won't coincide with when the pests show up in full force, and this will help avoid the worst onslaught.
- Start looking for any evidence of pest damage at pest-emergence times so you can nip pests in the bud before they become a problem (see *PESTS, VEGETABLE PLANT, emergence times for*).
- Give the pests a strong spray of water from the garden hose, hitting both sides of the leaves. Repeat every few days.
- Pick the pests off by hand or with tweezers, knock them from plants, or gather them with a net or a handheld vacuum, and then put them in a container filled with soapy water. Do this in the early morning, when the bugs are sluggish.
- Use horticultural oil spray for control of the adult and egg stages of many insects, including aphids, mites, beetles, leaf miners, caterpillars, corn earworms, thrips, leafhoppers, whiteflies, scale, and mealybugs. Ultra-refined oil is less temperature-specific than dormant oil spray,

but never use it within a month before or after spraying with sulfur (see *HORTICULTURAL OIL SPRAY*).

- Use insecticidal soap for aphids, mealybugs, spider mites, scale, thrips, whiteflies, leafhoppers, grasshoppers, earwigs, cucumber beetles, squash borers, psyllas, spittlebugs, and rose slugs (see *INSECTICIDAL SOAP*).
- Use sticky insect traps for whiteflies, fruit flies, young leafhoppers, flea beetles, cucumber beetles, squash bugs, fungus gnats, male winged mealy-bugs, leaf miners, and winged aphids (see *STICKY INSECT TRAPS, to make*).
- Use BT (*Bacillus thuringiensis*) for corn earworms, budworms, cabbage worms and loopers, and caterpillars in general. BT are organically approved, naturally occurring bacteria found in soil that are not harmful to beneficial insects, other creatures, or humans. BT only remains active for a day or two and can be used right up to the day of harvest. Spray weekly in early morning or evening. BT appears to works best on young larvae.
- Add one tablespoon of molasses per gallon of BT solution to encourage insects to feed on sprayed foliage. This is especially useful for large-scale spraying of fruit trees. Or use half the amount of insecticide called for by adding one tablespoon of molasses per one gallon of solution. The molasses increases the adhesiveness.

***PESTS, VEGETABLE PLANT, emergence times for***

- Contact your local extension office to find out when various pests become active in your area. The dates they emerge can vary from year to year, depending on the weather.

***PETUNIAS, hints on***

- Chill petunia seeds before sowing: Place them in a resealable plastic bag with some coarse sand or vermiculite, and put the bag in the refrigerator for three weeks. Sprinkle the seeds, along with the sand or vermiculite, over the seedbed. The small seeds will be distributed more evenly with the addition of the filler.

- Cut back the longest petunia stems to an inch or so from the ground after several weeks of heavy bloom, or when they grow tall and straggly. Then fertilize, and water well. This will result in compact new growth and prolong the flowering season.

***PITS AND CITRUS SEEDS, to germinate***

- Put slow-germinating date pits or citrus seeds in the soil of a potted plant. That way you won't have to remember to keep them moist. This works best with plants that need constant moisture. Make sure date pits are from unpasteurized dates.

***PLANT MARKERS, to make***

- Make plant markers using bleach bottles or plastic milk bottles, waxed milk cartons, large yogurt containers, aluminum dinner containers, or other foil containers. Cut them into ten-by-one-inch strips. Use an indelible marker, crayon, or grease pencil to write the appropriate information on the blank side of each strip.
- Use twigs with shaved off areas to make markers.
- Save Popsicle or ice-cream sticks and disposable plastic knives to use as markers.
- Cut pieces of old mini blinds so that one end is pointed and slides into the ground or seed tray.
- Coat plant markers with clear fingernail polish to make them more weatherproof.

***PLANT POTS;*** see *CONTAINERS FOR PLANTS, hints on; BUCKETS, hints on*

***PLANT TIES, to make***

- Cut lengthwise strips from old pantyhose, and use the strips to tie tomatoes and other plants to poles. The pantyhose stretches, so it won't cut into the stalks as they grow.

- Tie up plants with old stretchy double-knit fabric; it has more give than cotton.
- Save the plastic ties that bind the newspaper, especially the Sunday edition.
- Slice up strips of dry-cleaning bags. They make almost-invisible ties.
- Use tape from broken cassette tapes as plant ties. The tape stretches and doesn't damage stems.
- Recycle the elastic from worn-out clothing.
- Keep plastic garbage ties and paper-coated ties from vegetables or produce to use as plant ties.

***PLANTER AND FLOWERPOT DRAINAGE HOLES, to cover***

- Use any of the following items to cover the drainage holes in planters and flowerpots: old sponges cut into squares, pieces of activated charcoal, pebbles, stones, Styrofoam packing pellets, broken pieces of clay pottery, nylon net or mosquito netting folded into several layers, well-washed fruit pits, cracked walnut shells, coffee filters, used dryer sheets, pieces of old nylon hose, or pieces of old screens.

***PLANTERS AND FLOWERPOTS, to disinfect for reuse***

- Scrub moss or algae off clay planters with a stiff brush or nylon scrubber. Use steel wool for mineral deposits.
- Scrub planters and flowerpots with hot sudsy water, and rinse them. Leave them in the sun for a few days. Or disinfect them with a solution of one cup of chlorine bleach to nine cups of water. Rinse the pots thoroughly, and let them air-dry. If possible, soak the pots overnight in the bleach-and-water solution.

***PLANTERS SUNK IN THE SOIL, hints on***

- Wrap planters in a double thickness of old nylon hose to keep slugs and insects from entering through the drainage holes.
- Place a piece of wire screening over the drainage holes.

### *PLANTERS WITHOUT DRAINAGE HOLES, hints on*

- Line the bottoms of planters without drainage holes with one inch or more of pebbles or Styrofoam packing pellets (or broken-up Styrofoam packing materials). Use more material for taller containers.
- Double-pot planters that don't have drainage holes: Grow plants in clay pots, and then set them inside larger planters without holes. Place a flat stone or a layer of sand in the bottom of the larger pots without holes.

### *PLANTING TIMES, hints on*

- Plant when all prospect of frost is over, plant heat-loving plants when the soil is warm enough, and plant according to the requirements of your geographical area. For example, in mild-winter areas, fall, rather than spring, is the time to plant peas, spinach, pansies, and other flowers and vegetables that prefer cool weather.

### *POTATOES, hints on*

- Avoid planting potatoes with pumpkins, squash, or cucumbers, which lower the potatoes' resistance to blight.
- Plant potatoes near bush beans. The beans will repel the Colorado potato beetle, and the potatoes will repel the Mexican bean beetle.
- Check the undersides of leaves for evidence of the Colorado beetle, which lays bright orange-yellow eggs in rows or small clusters. Knock the beetles and their larvae into a bucket of soapy water, or shake the plant, and they will fall off. Spray the plants after dark with neem-oil, and mulch them with straw. The beetles don't like crossing over it.
- Control wireworms with bait. Bury some peeled and sliced potatoes about one and a half inches down around affected crops. Put a stick through each potato slice, and let the stick protrude out of the soil. Every couple of days, pull out the slices and wireworms, and replace the traps with new ones.
- Save space by planting potatoes in a circular compost bin that has two-and-a-half-inch holes in the sides. Or plant the seeds in a burlap bag. Fill

the bag with soil, and tie the end closed with wire or rope. Lay the bag down on its side, and cut small holes in the middle to plant the potatoes. Place the bag in a sunny spot, and keep the soil moist. When you're ready to harvest, cut the bag open, and gather the potatoes.

- Experiment with growing potatoes by using some store-bought ones. Spread some healthy-looking specimens in a shallow box, and put it in a warm room near a sunny window. In a few weeks, when the potatoes sprout, cut them into sections so that each piece has two or three buds. Let them sit for a day or two in a warm place to dry, and then plant them, cut sides down, in a six- to eight-inch trench. Cover the trench with soil, and then water it.
- Harvest early potatoes when the vines are just beginning to die, and harvest mature potatoes when the vines die back. Dry the potatoes for a day in the shade so that they will keep better.

***POTTING MIX, DRIED OUT, hint on***

- Add one or two drops of mild liquid dish soap to a gallon of so of water, and then soak the unused mix thoroughly. The soap helps rewet difficult-to-wet potting mixes.

**PROPOGATING CONTAINERS** see *SEEDS STARTED INDOORS, containers for*

***PRUNING, to protect your hands while***

- Use a spring-type clothespin, pliers, or kitchen or barbecue tongs to hold on to the thorny and prickly stems of roses and berries while you prune them.
- Put a pair of long oven mitts over gardening gloves when pruning roses or raspberry bushes.

***PUMPKINS, hints on;*** see also *SQUASH, hints on*

- Plant bush pumpkins if you're short on space.

- Avoid planting pumpkins near squash to avoid cross-pollination. Pumpkins and squashes also share the same pests and diseases.
- Grow an extra-large pumpkin by choosing a giant variety such as big Max or Atlantic giant. Leave only one or two pumpkins on each plant, and snap off all the other blossoms when they start to form. Allow a few fruits to reach grapefruit size, and then choose the best one—the one closest to the roots, with the thickest stem and best color or size and shape—and clip off all the others.
- Repel squash bugs by planting radishes and French marigolds. If you do see evidence of squash bugs, water all along the stalks to bring them out, and then pick them off, and drop them into a bucket of soapy water.
- Customize a jack-o'-lantern by writing a child's name on the pumpkin when it's about the size of a softball. Use a ballpoint pen, and lightly break through the skin while writing; this won't damage the pumpkin. As the pumpkin expands, so does the name.

***PUSSY WILLOWS, hints on***

- Cut some pussy willow branches in late winter, and place them in water. Change the water every day or so, or add a little charcoal to the water to keep it fresh. When the cuttings send out shoots, plant them outdoors.
- Plant pussy willows in a wet area of the garden. This will also help to dry out the wet area.

# R

***RABBITS, to repel***

- Break corncobs in half, soak them in white vinegar for five or six minutes, and then place them around the plants and vegetables. Repeat at two-week intervals. Keep the soaking vinegar in a covered container for repeated use.
- Edge the garden with dried sulfur from a garden- or farm-supply store.
- Erect a two-foot wire-mesh fence with a mesh of one inch or smaller.

***RACCOONS, to repel;*** see also *ANIMALS, to deter; CORN, to protect from raccoons*

- Plant winter squash or pumpkins around the perimeter of the property. Raccoons don't like crossing over the prickly leaves that scratch their bellies.
- Prevent the raccoons from eating birdseed from the feeder by hanging it on a wire between trees or on a baffled pole.
- Call your local animal-control center. For a fee, they will set up traps and return to pick up the raccoons.

***RASPBERRIES, hints on***

- Plant several different varieties of raspberries to stretch the harvest season.
- Train the vines on two horizontal wires for easier handling. Have the upper wire four to six feet above the ground and the lower one two and a half to three feet above the ground.
- Protect the fruit from birds by covering the bushes with old nylon curtains, cheesecloths, or nylon nets.
- Pick the berries every day during hot weather, or they'll become overripe.

- Protect your hands while pruning by wearing long oven mitts or a pair of heavy old leather gloves. Or put a pair of long wool socks over your arms. Cut a large hole and a small hole in the foot of each sock so you can slip your thumb and finger through. Use pliers or barbecue tongs to hold the branches.

***RED SPIDER MITES, to control***

- Remove red spider mites with a strong spray of water from the garden hose, hitting both sides of the leaves. Repeat every couple of days.
- Spray the mites with pepper spray. Spot spray first to see how the plant reacts; then apply the spray three times a week until you see results (see *PEPPER SPRAY*).

***RHUBARB, hints on***

- Avoid planting rhubarb in the shade; otherwise red-stemmed varieties will remain green.
- Confine rhubarb to its own area of a flower bed if you're short on space. It's colorful and will blend in well.
- Wait until two years after planting before you begin to pull stalks for eating. Select those that are at least one inch thick, and snap them off near the base. (Cutting stalks with a knife will leave stubs that might decay.)
- Harvest rhubarb only in early spring for one or two months (or until slender leaf stalks appear), and always leave at least half the plant intact. A summer's growth is needed to build up the plant crowns for the following year.
- Remove the flower stalks as soon as they appear so all the plant's energy is directed toward the edible stalks.
- Water plants well for a day or two before picking to ensure tender stalks.
- Make sure when dividing established plants that each piece has two buds on top or a piece of root. Position the new root so that the buds are just emerging from the soil.

- Freeze any excess rhubarb by cutting the washed and dried fruit into one-inch chunks, freezing them on a parchment-lined baking sheet until solid, and then packing them into freezer bags. The fruit will keep for up to nine months.

***ROOF DRAIN SPOUT, hints on***

- Place a large flat stone or garden stepping-stone beneath the drain spout to avoid soil erosion when it rains.
- Plant mint, marsh marigold, yellow flag, cardinal flower, and other thirsty plants next to the drain spout.
- Buy a Rain Drain or Drain Away, which automatically unwinds when it rains and gently releases water away from the house.

***ROSES, hints on***

- Place fish scraps in the bottom of the hole when planting roses, or dig the scraps into the soil close to an established rosebush. Fish is rich in phosphorus and potassium, plus iron. (Ask your fish merchant for free fish scraps.)
- Dress up the legs of rose bushes with skirts of companion plants planted at their bases. Good choices are catmint, lamb's ears, lobelia, Santa Barbara daisy, scabiosa (a.k.a. pincushion flower), sweet alyssum, or vinca. Broadcast evenly so the plants form a carpet under the roses. (The lower portions of rose bushes are not particularly attractive.) An added bonus of catmint is that it also repels aphids.
- Hose down the rose plants or give them an occasional overhead watering to remove dust from the foliage. This also controls aphids and spider mites and aids in controlling mildew by killing the spores. Do this early in the day to give foliage a chance to dry off. Occasionally, also give the roses a cleaning with soap and water to discourage insects and disease. Use one-quarter teaspoon of mild liquid dish soap to one gallon of water.

- Pull out sucker growth rather than cutting it. Suckers, which grow from the base rather than the bud union, have long, narrow, light-green leaves and are usually thornier than regular budded varieties.
- Feed the roses after pruning and spraying with dormant oil with a rose food formula developed by growers of old heritage varieties. This rose food is especially beneficial for alkaline soils and makes enough for one rose bush: Combine one cup of gypsum with one tablespoon each of soil sulfur, chelated iron, and Epsom salt. Mix the ingredients together, rake them into the soil around the base of the bush, and water.
- Wear leather or gauntlet gloves or long oven mitts when pruning in order to protect your hands.

***ROSES, to control pests and disease on***

- Interplant garlic, onions, chives, and shallots with roses. All members of the onion family help protect roses from mildew and aphids. Or bury three or four garlic cloves two inches deep around each rosebush. Garlic is a known accumulator of sulfur, which is effective against black spot.
- Grow tomatoes near or among roses. Studies have shown that tomatoes grown near or among roses consistently eliminate black spot on the roses.
- Plant a white geranium in the rose bed as a lure for Japanese beetles. Knock the beetles into a bucket of soapy water during early morning.

***ROW COVERS, hints on***

- Use fine netting or mosquito netting, a double layer of dry-cleaning bags, or plastic packaging from furniture, appliances, or mattresses to serve as material for the row covers.
- Check out thrift stores and garage sales for sheer nylon or polyester curtains in eight-foot lengths and extra-wide widths to use as material for the row covers.
- Support the row covers by draping them over hoops or supports made from wire coat hangers; eight-, nine-, or ten-gauge wire; bamboo; sticks; or PVC pipe. Secure the edges of the row covers with soil, mulch, rocks,

cans filled with stones or sand, or two-by-fours, or anchor the edges in place with stakes or U-shaped pins.

- Sew rips in material; fix holes and rips in plastic with duct or polyethylene tape placed on both sides of the plastic.
- Cut up any row covers that are beyond repair, and use them as wraps to protect ripening fruit.
- Remove the covers on beans, cucumbers, eggplants, melons, peppers, squashes, and tomatoes when flowers open so that pollinating insects can reach the blossoms.
- Leave the covers on carrots, onions, potatoes, and turnips all season if desired.
- Wash dirty row covers before storing for the season. Hang them on the clothesline, and rinse them gently with a hose. Or give them a good soaking in warm soapy water, then rinse, and air-dry.

# S

***SALAD GREENS, hints on;*** see also *LETTUCE, hints on*

- Salad greens should be given one inch of water every week (or enough to provide adequate moisture); feed them with liquid seaweed emulsion every two weeks, and give the plants shade when required. Rapid growth is the formula for succulent greens of maximum quality.
- Arugula is best used when its leaves are no more than two to three inches tall; otherwise they become bitter. Pull the plants before they go to seed; arugula seeds prolifically.
- Asian mustards and kale are best harvested by thinning out the baby greens as the plants mature. Another harvesting option is to cut the plant right back to within one inch of the ground; the crown will sprout new leaves for a second harvest.
- Endive makes a great lettuce substitute. It is more heat and cold resistant than lettuce, although you generally need to blanch it to lessen its bitterness and improve its flavor (see *ENDIVE [ESCAROLE], hints on*).
- Kale can be cut for salad when the leaves are three to five inches long. The plant grows well in cold weather and is sweetened after frost. When growth slows down and the outer leaves are tough, use the tender inner core for salad.
- Kohlrabi greens are best harvested for salad when the leaves are young and succulent. Kohlrabi is more tolerant of heat than other cool-season crops and can thrive in almost any kind of soil or temperature.
- Malabar nightshade is a tasty tropical salad green that thrives in hot weather and, with its red-veined leaves and black berries, makes a showy plant for edible landscaping.

- Mesclun is best harvested when the leaves are three inches to six inches long. Three inches is considered ideal for the tastiest leaves (see *MESCLUN, hints on*).
- Mustard greens are good in salads and grow to maturity fast. Harvest the young thinnings, and pick the leaves before they mature. The best-flavored ones are those that are four to five inches long. Keep the plants cut back to hold off flowering.
- New Zealand spinach is more heat tolerant than proper spinach and a good salad standby during warmer months. Cut three- to four-inch lengths from the tender growing tips, and use only the young leaves.
- Radicchio adds color and zip to green salads. This red-leafed chicory is drought tolerant and can survive temperatures down to 10°F (although it requires four hours of direct sun each day). Cut the tender leaves throughout the growing season.
- Spinach is best harvested for salad when the leaves are large enough to be used, about three to four inches long. Pinch or cut them individually from the plant, and pick them down to about the six centermost leaves. Take care not to harm the growing tip so that leaf production can continue. Plant closer together than usual, and harvest the thinnings to make room for other plants.
- Swiss chard's young tender leaves are best for salads. Harvest the outer leaves every few days.
- Beet and carrot thinnings (tops and roots) make excellent additions to salads. You can also harvest one-third of the tender beet greens without harming the roots.
- Excess salad greens can be served sautéed, stir-fried, or steamed.

### *SAP ON HANDS OR TOOLS, to remove*

- Rub the sap with vegetable oil, baby oil, or shortening, and then dry with paper towels. For hands, finish up by washing with soap and water.

- For hands and clothing, rub the sap with an alcohol-based hand sanitizer, and wipe it off with a piece of paper towel.

**SCOOP, *to make***

- Use an empty washed-out plastic bleach, milk, or orange-juice container: Cut off the bottom and cut out one side (up to the handle). Use the top with the cap attached for scooping compost or potting soil.

**SEASHELLS, *uses for***

- Use seashells as a decorative mulch in the garden. Use small ones as is, and break up larger shells with a hammer.
- Use shells to provide humidity for houseplants. Spread them on trays with water, and set houseplants on top.
- Put shells in the bottoms of pots to line the drainage holes.
- Place shells in the bottom of a vase of low-stemmed flowers to add leverage.

**SEAWEED TEA, *to make;*** see also *FERTILIZER, hints on*

- Pour boiling water into a bucket of well-rinsed chopped kelp, and let it stand several hours or overnight. Strain the liquid, and spray it on the plants. According to research, seaweed is not only the most complete food material you can feed a plant, but it also improves the plant's overall health.

**SEED GERMINATION, *requirements for***

- Read directions on the seed package for planting instructions. The majority of seeds require darkness to germinate, but some seeds germinate better when exposed to light. Sow the following seeds on the surface, and don't cover them with soil mix: ageratum, begonia, cineraria, coleus, dill, feverfew, flowering tobacco, German violet, gloxinia, impatiens, lettuce, lobelia, petunia, primrose, salvia, Shasta daisy, snapdragon, stock, strawflower, sweet alyssum, and yarrow. (It is usually the finer seeds that require light to germinate.) Place the seed container in bright but indirect light until germination occurs.

***SEED TAPES, to make***

- Dissolve one teaspoon of cornstarch in one-third cup of cold water; then cook the mixture over low heat, stirring constantly, until the mixture boils. Remove it from the heat, and let it cool. Cut long strips of paper toweling two to two-and-a-half inches wide, and place dots of the cornstarch mixture along the lengths of the strips. Place one seed on each dot, and then let the gel dry. Roll up the seed tapes, and store them in a plastic bag. Put them in the garden before or after the seeds sprout. The paper and cornstarch mixture will decompose in the soil.
- Make a paste with 1 tablespoon flour and 2 tablespoons cool water, then use it to paint a dot on one side of toilet paper strips. Place a seed on each dot, let the paste dry, and then fold over the other side of the toilet paper.

***SEEDLINGS, to transplant;*** see also *TRANSPLANTING IN HOT WEATHER, hints on; SEEDLINGS IN THE GARDEN, to shade; SEEDLINGS IN THE GARDEN, to protect from cutworms*

- Use a pair of small manicure scissors to cut superfluous seedlings off at soil level.
- Use a metal nail file, a kitchen or putty knife, or a small spatula to separate seedlings at transplanting times.
- Help small seedlings get established quickly by burying their stems up to the top pair of leaves.
- Help leggy or spindly transplants develop stronger root systems by stripping off the lower leaves and laying the stems in an angled trench in the garden. The plants will form roots along the buried stems. This procedure is also beneficial for tomato transplants.

***SEEDLINGS IN THE GARDEN, to protect from cutworms***

- Lay a mulch of crushed eggshells, wood ashes, or oak leaves around new plants to protect them from cutworms.
- Apply cutworm collars when setting the seedlings in the ground. Press the collars into the soil one inch, leaving one or two inches above the soil (see *CUTWORM COLLARS, to make*).
- Dig around a plant destroyed by cutworms; usually the cutworms are just under the soil nearby. Drop them into a container of hot soapy water.

***SEEDLINGS IN THE GARDEN, to shade***

- Prop up a board, window screen, or umbrella next to the seedlings for shade.
- Place a berry basket over each individual seedling.
- Put a milk crate or mesh laundry basket over a group of seedlings.
- Suspend cheesecloth, old sheer curtains, nylon netting, or shade cloth over the seedlings. Tie the shade material to stakes.
- Use an old sheet as a canopy for a large area. Fasten the corners to six- to eight-foot stakes.
- Make long shallow tents of window screening by bending the screens in half lengthwise. Place the tent over the seedlings, and cover the edges of the tent with soil.
- Place light lawn chairs over seedlings to provide temporary shade.

***SEEDLINGS IN THE GARDEN, to water***

- Spray seedlings with a fine mist from the garden hose, or use a watering can with a rose nozzle.
- Place a soil-soaker hose next to the seedlings. (Canvas or plastic soaker hoses can be purchased from a garden-supply store.) Or punch small holes in an old plastic hose, and lay it on the soil between the rows, holes down.
- Slip a cotton glove or sock or piece of burlap over the hose nozzle, and fasten it securely with string or a rubber band. Lay the covered nozzle next to the seedbed.

***SEEDS SAVED FROM PREVIOUS YEARS OR PREVIOUS CROPS, to test***

- Test seeds before planting to see whether they are still viable by putting about ten seeds in a glass of water and letting them soak. Dry, useless seeds will float, while viable seeds will sink. To determine seeds' germination rate, place ten to twenty seeds on one side of a wet paper towel, and fold over the other side to cover the seeds. Place the paper towel in a plastic bag to keep it moist, and then put it in a warm place. Record the date, and check the seeds regularly. Germination may take a few days to a few weeks. If at least seven out of ten seeds sprout, the seeds are viable. If only a few seeds sprout, either discard them or sow them extra generously.

***SEEDS STARTED INDOORS, containers for***

- Inquire at nurseries or plant growers to see whether you can obtain their surplus flats. They may have empty flats left over from shipments, or they may sell you one or two for very little money. They may also have two-inch pots available that are left over from transplanting plants into larger containers.
- Check with landscape contractors—they often have leftover flats from ground-cover plantings, plus other pots and containers.
- For small seeds, use a cardboard or Styrofoam egg carton with the top cut off. Poke holes in the bottom, and put in little pieces of eggshell to make removal of the seedlings and soil easier. Plant the whole cardboard egg carton directly in the soil. Or cut it into twelve sections, and plant them individually. (The cardboard will decompose in the soil.)
- For large seeds, use paper or Styrofoam drinking cups, empty K-cups, single-serve yogurt cups or other small food containers, or half-pint or quart-size milk cartons with the tops cut off. Poke holes in the bottom for drainage using a heated, straightened-out metal coat hanger, a large nail, or a Philips-head screwdriver. Use the cups and containers for cutworm protectors when replanting. Remove the bottoms, and sink them into the soil.

- For small flats, use disposable foil containers or shallow cardboard boxes completely lined with aluminum foil. Poke holes in the bottom.
- Line strawberry baskets or cardboard mushroom or berry containers with thin layers of newspaper. When the plants are ready for transplanting, slip the newspaper-enclosed soil from the baskets, and plant without removing the newspaper. The paper will decompose in the soil.
- For a greenhouse-like atmosphere, use clear plastic food containers, such as clamshells or carryout salad containers. Poke a few drainage holes in the bottom. Close the lid with a rubber band, if necessary. This will provide a humid environment for the seeds.
- For plants that don't like their roots disturbed, cut tubes from paper-towels, wrapping-paper, or toilet-paper rolls into one-and-a-half to two-inch sections. Stand the cylinders upright side by side in a container, fill the pots with dampened potting mix, and water until the cardboard is saturated or turns dark. Plant one seed in each container. When the seedling are ready to plant, sow them directly in the ground. These pots are good for vegetables such as beans, corn, cucumbers, and squash.
- Buy peat pots, which make good starting containers for plants that don't like their roots disturbed. The pots can be planted directly in the ground.
- Sow more than one seed in each peat pot or other container. Upon germination, cut off any extra seedlings at the soil level.
- Reuse plastic containers a few times by washing them thoroughly with soap and water and disinfecting them in a solution of one cup of chlorine bleach to nine cups of water. Cardboard milk cartons, however, can only be used once.

***SEEDS STARTED INDOORS, starter mix for***

- Use fine potting soil, vermiculite, or a soilless mix. Put the chosen starter in a plastic bag, filling the bag half full; then dampen the mix with warm water at the rate of one part water to four parts starter mix. Mix well by squeezing the bag gently until the water is evenly distributed. Fill the plant container to within one-quarter inch from the top, and then loosely pack down the starter mix. If the container is too deep, you can put a layer of perlite or Styrofoam in the bottom to reduce the amount of starter mix required.

**SEEDS STARTED INDOORS, *to avoid dampening off (a plant disease occurring in excessively damp conditions)***

- Fill flats and containers almost to the top with seed-starting mix. This will allow airflow across the seedlings and avoid dampening off.
- Avoid seed-starting mixes that contain fertilizer. Apply diluted fish fertilizer or seaweed extract when the seedlings have their first sets of true leaves.
- Let the soil dry out between watering, and water from the bottom.

**SEEDS STARTED INDOORS, *to boost the lighting for***

- Line a shelf or windowsill with aluminum foil, or surround the seedlings on the windowsill with foil-covered cardboard to create reflected light. Cut the bottom and one long side off a cardboard box, and attach aluminum foil to the inside of the remaining three sides. Place the reflector box facing a window with the open side toward the window, and set the plants inside.
- Prop up an old mirror behind the seedlings to boost the amount of light they receive.
- Put the seedlings in a southern window if you live in a colder zone and in an east-facing window if you live in the South. Seedlings need light, not necessarily heat.

**SEEDS STARTED INDOORS, *to grow under lights***

- Have the lights on your seedlings for fourteen to sixteen hours a day every day, and, if possible, use an automatic timer.
- Keep the seedlings three inches away from the lights when the seeds are sprouted, four inches away when they have their first true leaves, and six inches away when they are a few inches tall.

**SEEDS STARTED INDOORS, *to water***

- Mist seeds gently with a spray bottle. Or place the pot or flat in a shallow pan containing water or a nutrient solution. Let the seedling pots soak up water until the top of the soil is moist; then remove and let drain.

**SEED-STARTER TRAYS** see *SEEDS STARTED INDOORS, containers for*

**SEEDS, *hints on***

- Save money on seeds by buying them from a nursery with good bulk-purchase prices and splitting them with neighbors.
- Buy seeds in spring for midseason or later planting. If you leave it for later, the varieties you want may have sold out.

**SEEDS, *to chill when required***

- Put a layer of moist potting soil on the bottom of a container such as a washed-out plastic lettuce or salad-mix container with a lid. Sow the seeds thickly, and cover them with more potting soil. Seal the container tightly, put it into a resealable plastic bag, and secure it with a rubber band. Leave the container at 60° to 65°F for three days; then refrigerate the seeds for the recommended period of time. When planting the seeds, sow them very close to the surface of the soil, and cover the soil with plastic wrap until the seeds germinate.

**SEEDS, *to save from the garden;*** see also *SEEDS, to store*

- Choose standard open-pollinated varieties of seeds, not hybrids; hybrid seeds may be sterile.
- Allow the plant to go to seed. (When the pods mature, they swell, turn brown, and dry out.) Hold a paper bag under the seed pod, and squeeze it so the seeds pop out. Label the bag, and store in an airtight container.
- Avoid losing any seeds by cutting the plant's stems near the base when the seeds are turning brown. Tie a bunch of stems together with a rubber band and put them in a paper bag, seed heads down. Close the bag with

a piece of string or a rubber band, and hang it in a cool, dry place. The seeds will dry and fall into the bottom of the bag.

- Pick vegetables whose flesh contains seeds, such as tomatoes, squashes, and cucumbers, when they are a little overripe. Remove the seeds from the flesh, and rinse them in water. Let the seeds dry for several days on newspaper or paper towels; then label them, and store them in an airtight container.

***SEEDS, to scarify (weaken the seed coating to speed up germination)***

- Cut or file through the edge of the seed at any spot other than the eye from which it will sprout. Hold onto the seed with pliers, nail clippers, or tweezers.
- Rub small seeds between two sheets of sandpaper or emery cloth.

***SEEDS, to set in rows***

- Use a broom or a rake handle to form a quarter-inch trench for your seeds, or use a piece of string to position the row.
- Use the perforated edges of continuous-form computer paper as a guide to sow seeds. The holes on computer paper are spaced a half-inch apart, so you can use the spacing you need.
- Place a piece of one-inch chicken wire over the soil, and use it as a guide to space the seeds.
- Mark the rows by planting radishes as row markers for slow-germinating crops. Plant a few of these fastest-growing seeds in the same row with other seeds.

***SEEDS, to sow in hot weather;*** see also *TRANSPLANTING IN HOT WEATHER, hints on*

- Soak seeds overnight, or place them in the refrigerator for a day or two before sowing them during hot weather.
- Plant seeds a little deeper than usual and in the shade of other flowers or vegetables. By the time the new plants need the space, the shade crops will have run their course.

- Place a board or a piece of plywood or carpet over the seeds until they germinate; then shade the seedlings until they are established (see *SEEDLINGS IN THE GARDEN, to shade*).

***SEEDS, to sow in the garden***

- Cover the seed row with strips of burlap, old sheets, or any porous material that will absorb moisture. Sprinkle it thoroughly every morning. Remove the cover as soon as the seeds sprout. Alternatively, cover the seedbed with a thin layer of dried grass clippings to prevent the soil from crusting.
- Increase the chances of germination in cool weather by placing clear polyethylene or plastic wrap over the newly planted row or bed and anchoring it with rocks or soil. The covering will speed up germination by warming the soil, keeping it moist, and preventing a crust from forming. It will also prevent birds from eating the seeds. Remove the covering as soon as seedlings emerge.

***SEEDS, to store;*** see also *SEEDS, to save from the garden*

- Store loose seeds in old prescription or pill bottles, plastic film canisters, sanitized herb bottles, or small jars with screw-top lids. Label the containers with peel-off labels, and enclose silica gel packets to retard moisture. Or put a little powdered milk in a piece of cheesecloth or facial tissue, twist it shut, and place it in the container with the seeds. Replace the powdered milk every six months.
- Keep seed packages in a small plastic box. Use index cards to sort the packages alphabetically, by planting season in weeks and months, or into categories such as greens, roots, flowers, and so forth. The box fits neatly on the refrigerator shelf if necessary and is readily accessible.
- Freeze the seeds in tightly sealed foil packages in a wide-mouth canning jar with a rubber seal. Short-lived seeds such as carrots, parsnips, onions, and leeks are good candidates for freezing.

**SEEDS, SMALL, to sow**

- Pour small seeds onto a plate, and pick up a seed with a moistened toothpick or pencil point; then drop the seed onto the soil surface. The seed will drop from the point when the toothpick or pencil touches the soil.
- Dip a child's small paintbrush into the seeds to pick up a few at a time, and then tap or stroke the bristles on the soil to release the seeds.
- Add a few teaspoons of coarse sand or dried coffee grounds to the seed packet. Hold the packet closed, and shake it to mix. Spread the mixture evenly over the soil surface.
- Sift a light layer of sand or flour over the surface of the seedling trays. You will be able to see where the seeds are placed.
- Fold a piece of paper in half, put the seeds into the fold, and gently tap the paper with your finger to scatter the seeds over the soil.

**SEEDS, TREE, to propagate** see *TREE SEEDS, to propagate*

**SHASTA DAISIES, hints on**

- Plant single Shasta daisies in full sun to have them flourish; plant the double varieties in partial shade to have them do their best.
- Use hedge clippers to shear off faded blooms when there are too many to deadhead individually.
- Dye cut blooms by dipping freshly cut flowers into water containing food coloring.

**SLUGS AND SNAILS, to control**

- Put out a board or shingle or inverted flowerpots where slugs and snails are likely to be at night. Or put out some spoiled lettuce, spinach, or cabbage leaves, or hollowed-out inverted citrus or melon halves. In the morning, lift the traps, and scrape the snails into a bucket of soapy water.

- Put used coffee grounds around acid-loving plants. Besides keeping slugs away, the grounds make great food for earthworms and may even help to repel some insects.
- Lay sandpaper around the plants snails favor, or make sandpaper collars. Cut a slit to the center of a circle of sandpaper and slip the collar around the plant at ground level. (Recycle used sandpaper discs from orbital sanders. Ask for used sandpaper from high-school woodworking shops or cabinetmakers.)
- Sprinkle coarse sand, gravel, sawdust, or garden-variety diatomaceous earth around each plant. (Obtain diatomaceous earth from a garden-supply store. Ask for sawdust at a lumberyard, firewood lot, cabinetmaker's shop, or high-school woodworking shop.)
- Put out empty tuna cans or jar lids containing beer (add a pinch of brewer's yeast for extra enticement). Or cut the bottoms off milk cartons to make shallow dishes one and a half inches deep. Sink them in the ground with their tops one-eighth inch above soil level; then put in some beer (or a mixture of equal parts beer and water sweetened with one teaspoon of sugar per one cup of water). Dispose of the snails by dropping them in a bucket of soapy water. Or if they've drowned, put them into the compost.
- Soak dry dog food in water until it's moist and mushy, and then put small clumps of it directly on the ground. Go out after dark with a flashlight, and dispose of the snails that have clustered around the bait.
- Edge the garden beds with copper tape or strips pressed about one inch in the soil. Or attach the tape to the top edges of raised beds with small tacks. Copper tape carries a mild electrical charge that is delivered by natural ground currents and is undetected by humans but is repellent to snails. Make sure there are no snails in the area before putting up the strips.

***SNAPDRAGONS, hints on***

- Avoid rust by
  1. planting rust-resistant snapdragon varieties;
  2. changing the planting location yearly; and

3. avoiding overhead watering. Rust is a common disease with snapdragons.

- Pinch off the tops of the plants when they are three to four inches tall to have them grow sturdily instead of lankily and to encourage more bloom spikes.
- Cut the plants back by one-third every ten days to two weeks so they keep producing.
- Cut the plants back halfway after they finish blooming, and then give them a dose of fertilizer. They should produce a second round of blooms.

***SOIL, to cover seeds***

- Use an old shaker container, colander, or kitchen sieve to sprinkle the soil over newly planted seeds in a thin layer.
- Nail a piece of plastic mesh to a frame for sprinkling soil over a large area. (Use a sixteenth-inch mesh.)

***SOIL, to sterilize for potting***

- Sterilize soil in the microwave by half filling a plastic bag with moistened garden soil. Use one part water to four parts soil. The soil should be damp but not wet. Seal the bag and make a hole in the top. Heat the bag on High for three minutes per quart of soil (or until it is steaming); then let the soil cool.
- Sterilize soil in the oven by spreading moistened soil in a shallow pan and baking it at 180°F for thirty minutes, or until a thermometer registers 180°F. Let the soil cool before using it. Baking the soil at a low temperature minimizes the odor.

***SOIL, to test***

- Call your local Department of Agriculture Cooperative Extension Service for information on where to send your soil for testing, or ask whether the local Extension Service will do it for you.

- Test soil pH quickly and inexpensively with blue litmus paper (obtainable at a pharmacy or science-supply store). Water the area, wait a few minutes, and then press the paper to the damp earth. Repeat the procedure wherever you want to test for acidity. If the paper turns pink, the soil is acidic. (You can also buy a soil-test kit from a garden-supply store for about ten dollars.)

### *SOIL, ACIDIC, to make more alkaline*

- To make acidic soil more alkaline, add ground limestone, dolomite, pulverized eggshells, clam shells, or oyster shells.

### *SOIL, ALKALINE, to make more acidic*

- To make alkaline soil more acidic, add soil sulfur, leaf mold, oak sawdust (add high-nitrogen fertilizer at the same time if nitrogen has not been added to the packaged product), tea leaves, or coffee grounds. (Inquire at local coffeehouses whether you can pick up their used grounds.)
- Add gypsum for heavy-clay alkaline soils. Besides reducing salinity (gypsum contains 22 percent calcium and 17 percent sulfur), gypsum binds organic matter to the soil, increases air circulation, and promotes drainage.
- Put a pinch of Epsom salt in the hole when planting acid-loving vegetables such as tomatoes, peppers, or eggplants. The magnesium helps set blossoms, and the sulfur adds acid.

### *SOIL, POORLY DRAINED, hints on*

- Plant blue and yellow flag irises, canna lilies, variegated sweet flags, elephant's ears, marsh marigolds, water lilies, wildflowers that thrive in bogs, and moisture-loving ferns.
- Plant a native pussy willow to help dry up the area. The roots are not as invasive as willows' roots.

***SOW BUGS, to control***

- Put out orange-rind halves overnight (leave a little of the pulp intact) to attract sow bugs. Next morning, dispose of the bugs in a container of soapy water.
- Place a clay flowerpot upside down wherever sow bugs are appearing. Let the pot sit for a couple of days; then remove the sow bugs from inside the pot, and drop them into a container of soapy water.

***SPACE-SAVING, hints on;*** see also *CONTAINERS FOR PLANTS, hints on; SUCCESSION PLANTING, hints on*

- Use hanging baskets or window boxes to grow herbs, cherry tomatoes, and other small vegetables to save space.
- Plant vegetables in pots and bushel baskets placed on the patio, beside a front door, along the driveway, or where the pavement meets the house or garage wall.
- Plant potatoes in a circular compost bin that has two-and-a-half-inch holes punched out of the sides.
- Grow root crops and potatoes in a burlap bag turned on its side and filled with soil (see *POTATOES, hints on*).
- Plant strawberries in a strawberry pot, an old gutter spout attached to a fence (seal the spout ends with tape), or window boxes.
- Train tomatoes, beans, peas, cucumbers, eggplants, melons, summer squashes, and small-fruit winter squashes to grow on a post, tepee, or trellis. Be sure to support heavy fruit with cloth slings. Erect the supports toward the back of the garden to hold the climbing plants.
- Plant a strip garden in the lawn for quick-growing greens or other vegetables. Broadcast annual rye anytime the strip is not in use to prevent the soil from washing out.
- Plant vines to grow on clothesline posts, downspouts, or chain-link fences. This will add color and fragrance and attract beneficial insects to the garden. Choose twining vines such as morning glory,

moonflower, wisteria, or honeysuckle, and keep them trimmed and within bounds.

- Plant vegetables in square grids or in raised beds rather than rows. These methods allow closer planting and easier weeding and watering, and the plants will produce more vegetables per area of garden space.
- Choose medium- and small-fruited varieties of tomatoes and peppers. The smaller the fruits, the more the plants tend to produce.
- Choose fast-growing crops and dwarf flowers and vegetables.
- Prune the roots and tops of container-grown ornamentals to keep them within bounds and avoid having to repot to larger containers. Prune in early spring when the plant is in active growth, and keep the roots and tops in balance. If you remove one-quarter of the roots, make sure you remove one-quarter of the top growth.

### *SPIDER MITES, to control*

- Keep spider mites under control by hosing them off of plants every few days. Spider mites reproduce rapidly, especially in dry, hot weather.
- Spray the mites with insecticidal soap or horticultural oil spray at five- to seven-day intervals until the mites are under control (see *INSECTICIDAL SOAP; HORTICULTURAL OIL SPRAY*).
- Combine two tablespoons of buttermilk (or regular milk that's been left out to sour), one-half cup of flour, and three quarts of water. Mix well to avoid lumps, strain the solution so it won't clog the sprayer, and then shake it each time you spray. Wash off any residue from the plants after it dries, and spray at seven-day intervals until the mites are gone.

### *SPINACH, hints on*

- Assist germination by soaking spinach seeds overnight. When planting in warm weather for a fall crop, soak the seeds for twenty-four hours. Another option is to freeze the seeds for a few days, and then moisten and refrigerate them between moist paper towels for a few more days before sowing.

- Plant the seeds directly in the ground and in the shade of tall crops; then put a piece of cardboard over the seeds to further hasten germination.
- Avoid planting spinach with potatoes; they are not beneficial companions. Beans and peas and cabbage are companions of choice.
- Cover the plants with row covers to protect them from leaf miners and flea beetles (see *ROW COVERS, hints on*).
- Plant radishes as a lure. Insects will go for the radish leaves rather than the spinach leaves.
- Harvest the outer spinach leaves when they are at least three to four inches long. If you pick just the outer leaves, the inner ones will become the next crop and extend the plant's productivity, but never take more than half of the plant at one time. Another option is to cut the whole plant back to one inch from the ground while its leaves are still small and tender; it will soon grow back.

***SPRAYS, ORGANIC GARDEN, to make;*** see *INSECTICIDAL SOAP; FUNGICIDE SPRAY; GARLIC-OIL SPRAY; HORTICULTURAL OIL SPRAY; and PEPPER SPRAY.*

***SPRINKLER HEAD OR HOSE NOZZLE, CLOGGED, to remedy***

- Remove the clogged head or nozzle, and rinse it; then use a toothpick to clean out the holes. If the nozzle is still clogged, soak it in white vinegar or a little calcium, lime, and rust remover, for one hour. Rinse before using.

***SQUASH, hints on***

- Group squash seeds together three or four inches apart, or plant them with corn; corn and squash are beneficial companions. Avoid planting squash with potatoes; potatoes have an adverse effect on squash.
- Put mulch, a board, or inverted plastic containers under winter squash. Doing so will prevent the fruit from rotting.
- Trellis the plants. Doing so will conserve garden space, keep the plants off the ground, maintain good air circulation, and prevent mildew.

Pinch off the growing tips when the vines grow to five feet to encourage fruit-bearing side shoots.

- Pick fruit when it's young and tender and before blossoms drop off the tip of the plant; otherwise the plant will stop producing.
- Let winter squash grow as long as you can; the longer it grows, the less water it will contain. Then let the squash sit for two to three weeks in a warm, dry place to cure. This will lower the moisture content, increase the sugar content, and improve the flavor. Acorn squash is the one exception. It does not improve with storage on or off the vine, so pick, and eat it when it is mature.
- Control powdery mildew by spraying every ten days with a milk fungicide spray (see *FUNGICIDE SPRAY, to make*).

### *SQUASH PESTS, to control*

- Cover young squash plants with light-fitting cheesecloth, mosquito netting, or nylon pantyhose as protection from squash bugs and cucumber beetles. Remove the covering before blossoms set so pollinating insects can pollinate the blossoms.
- Loosely wrap the base of the plant with a piece of aluminum foil to protect it from squash borers. Do this as close to the soil as possible. If you do see a hole in the stem, slit the stem where the hole is, and remove the borer; then cover the area with soil or mulch.
- Put flat boards around the plants to attract squash bugs; then dispose of the bugs.
- Plant radish seeds to deter cucumber beetles, squash bugs, and borers. Leave the radishes in place all season for full effectiveness. French marigolds also act as a repellent.

### *SQUIRRELS, to keep away from bird feeders*

- Wrap transparent tape or duct tape, sticky side out, around the bird feeder's chain.

- Spray the chain and top of the bird feeder with nonstick cooking spray, and then sprinkle the sprayed area with cayenne pepper. (Buy the pepper in bulk.)
- Mount the bird feeder on a pole, and coat the pole with heavy all-purpose grease, such as axle grease, petroleum jelly, or Tanglefoot.
- Purchase a Squirrel Baffle for under ten dollars if all else fails.

***STAKES, hints on***

- Use the ribs from an old umbrella to stake small plants; the ribs are strong and can hardly be seen.
- Save old broom and mop handles to use as stakes for small trees or to provide support for climbing plants. Press them about one foot into the soil to anchor them properly.
- Bend wire clothes hangers into stakes. Straighten the hooks completely, and then pull the bottom portions out to form large diamond shapes. Put the straightened hooks into the ground, and use the diamond portions to help support the plants.
- Use twigs and pruned branches from fruit trees to prop up plants.

***STICKY INSECT TRAPS, to make***

- Use yellow items (such as coffee-can lids, yellow paper cups, or rectangles cut from bright-yellow plastic bottles): Smear them with a sticky coating (such as axle grease, petroleum jelly, molasses, or Tanglefoot). Or wrap each item in plastic wrap, and apply the coating to the plastic. Tack the traps on short stakes, and position them two to four inches above the ground for leafhoppers, or two to six inches for most other insects. Remove the sticky coating periodically, and reapply it. Alternatively, smear the sticky substance on a yellow detergent bottle, and set it near the infested plant.

***STINK BUGS, to control***

- Knock or shake stink bugs from the plants into a container of soapy water.

- Put out sticky insect traps (see *STICKY INSECT TRAPS, to make*).
- Plant chrysanthemums, lavenders, marigolds, and sunflowers as traps. Doing so will make the stink bugs easier to locate.

***STRAWBERRIES, hints on***

- Plant strawberries in raised beds for better air circulation and drainage.
- Mulch strawberries with newspaper or black plastic to keep the ground cool and the soil moist. (Plant the strawberries through slits in the plastic. For aesthetic purposes, cover the plastic with a thin layer of straw.)
- Ensure a lush second-year crop by pinching off all the blossoms after planting and all the blossoms during the first year.
- Dig out the oldest crowns every year or two so the younger ones can take over and keep the bed vigorous.
- Prevent birds from eating the ripening strawberries by covering the strawberries with nylon net or mesh bags (from onions or fruit) that you've opened up and stitched together. Suspend the protection above the strawberries so the birds can't poke their beaks through to reach the fruit.
- Keep the birds from eating the fruit by sticking a metallic pinwheel in the middle of the bed. (Buy pinwheels at a dollar store.) Birds don't like the sound or movement pinwheels make. (Don't depend on this if there is no air movement.) Another deterrent is to put small red-painted stones or pebbles around the strawberries. The birds will mistake them for strawberries.
- Conserve space in the garden by planting the strawberries in planters, window boxes, or an old gutter spout with holes drilled in the bottom, the ends taped up, and attached to a fence.

***SUCCESSION PLANTING, hints on***

- Plant fast-growing vegetables between slower-growing kinds to get the most out of your garden space. For instance, try putting beets between the peppers and tomatoes or lettuce between tomatoes and beans.

They'll be ready to harvest before the larger, slower crops need the room. Plant everything in different parts of the vegetable garden so diseases don't build up in the soil.

- Interplant cabbages, cauliflowers, lettuces, and bush beans with cucumbers. The vegetables should be producing around the time the cucumbers come up and need the space.
- Plant spinach, lettuce, or Chinese cabbage at the base of trellised peas, where the greens will benefit from the shade and wind protection.
- Follow spinach or cucumbers or other heavy feeders with beans. The beans will return nitrogen to the soil.
- Plant turnips and rutabagas between cabbages. They are all cool-season vegetables that need a steady supply of water.
- Make a new sowing every ten days to two weeks. Mix early, midseason, and late-maturing plant varieties.

***SULFUR SPRAY, hints on***

- Avoid spraying sulfur directly on the soil. A high concentration of sulfur can damage soil microbes.
- Do not use sulfur spray in hot weather over 80°F, or the sulfur will damage the plants.
- Never use oil sprays within a month before or after spraying with sulfur or compounds containing sulfur.
- Avoid getting the sulfur spray on the side of the house or a painted fence. Sulfur discolors paint.

***SUNFLOWERS, hints on***

- Soak the hard-shelled sunflower seeds for half a minute in boiling-hot water, and then plant them while they're still hot and moist. (Remove the boiling water from the heat before adding the seeds.) Or soak the seeds overnight in lukewarm water, and then plant them.
- Plant tall varieties of sunflowers on the north side of the garden so they don't shade other crops.

- Plant the small sunflower varieties six inches apart if you want them for cutting. Close and crowded planting produces lots of small flowers on longer stems.
- Pinch off the first bud on multiflowered or branching varieties. The plants will bush out and produce more blooms.
- Grow the dwarf varieties in blocks for the best display, and keep the flowers picked to prolong the blooming period.
- Cut the flowers in early morning or late evening, when the buds are just beginning to open; then put them immediately into lukewarm water, and change the water daily. Adding a few drops of chlorine bleach to the water will prevent it from turning murky.
- Plant multiflowered types for drying, especially Sunbright. Cut the flowers when the heads sag down, leaving enough stem to hang them upside down, and then air-dry them for two to three weeks in a dry, airy place.
- Keep birds from eating the ripening seeds of tall varieties by covering the flowers with sections of mesh bags that onions and fruit come in. Keep the bags in place until the seeds are dry enough to harvest.
- Harvest the seeds by leaving the sunflowers in place until the plant turns yellow and withers and the back of the sunflower turns completely brown. Remove the seeds with a stiff brush or by rubbing two of the heads together.

### *SWEET ALYSSUM, hint on*

- Use garden shears to clip off all the old flower heads when the sweet alyssum plants stop blooming. This will spur the plants into another flowering cycle.

### *SWEET PEAS, hints on*

- Soak sweet pea seeds in water for twenty-four hours before planting; then push each seed one inch into the soil or potting mix.
- Keep the seeds away from children; they are poisonous.

- Cut sweet peas for indoor arrangements when the flowers just open. Stand them in a cool mixture of eight drops rubbing alcohol per quart of water. To have the blooms last even longer, soak them in floral preservative and water in a cool place overnight before arranging.

***SWISS CHARD, hints on***

- Soak seeds overnight to hasten germination.
- Interplant Swiss chard with onions, which are friends, but not with pole beans, which have an adverse effect on Swiss chard.
- Use row covers to protect the plants from leaf miners, and leave the covers on all season if needed.
- Harvest Swiss chard every few days by breaking off the outer leaves from the plants when they are at least five inches long. (Using a knife sometimes causes the plants to "bleed.") Don't let the old leaves remain on the plant, or the plant will stop producing fresh leaves. Another harvesting option is to cut off the whole plant a couple of inches above the root crown. Fertilize the plant, and it will grow back in about four to six weeks.
- Use the young tender leaves as an addition to salad. Cook the outer leaves as you would spinach or other greens, and cook the stalks as you would asparagus or bok choy.
- Leave Swiss chard in the ground during the winter; the foliage will die back, but the plant will make rapid regrowth during early spring.

# T

## *THRIPS, to control*

- Spray them with insecticidal soap (see *INSECTICIDAL SOAP*).
- Set out blue sticky traps: Smear four-by-six-inch pieces of blue plastic or cardboard with molasses or heavy motor oil, and place them on or a few inches above the infested plant. Remove the thrips with vegetable oil periodically, and renew the coating.

## *TOADSTOOLS, to remove*

- Rake persistent toadstools, and then spray them with a solution of two tablespoons of baking soda and one teaspoon of liquid soap mixed with one gallon of water.

## *TOADS, to attract to the garden*

- Partially bury an old clay flowerpot on its side, and put a shallow dish or pie plate filled with water next to it to attract toads. Replace the water daily to discourage mosquitoes. Or break a small two- to three-inch hole in the side of the pot to make a doorway, and turn the pot upside down. Toads, which help keep the garden free from pests, might set up residence in this toad sanctuary.

## *TOMATOES, hints on*

- Plant tomatoes near asparagus, basil, carrots, celery, chives, and parsley, which are all beneficial companions.
- Avoid planting tomatoes with corn, kohlrabi, or potatoes. These plants have an allelopathic effect on tomatoes, which retards their growth. Also, avoid planting tomatoes where potatoes, peppers, or eggplant grew the

previous year. They are all members of the nightshade family and are susceptible to the same pests and diseases

- In cool climates, put down a sheet of black plastic a few weeks before planting tomatoes. The plastic will warm up the soil and keep it warm. Then use it as a mulch once the plants are in the ground. Or put down clear plastic sheeting for a week or so; then remove it before planting. Clear plastic will warm the soil by eight to ten degrees.
- Plant tomatoes where summers are cool in an area that slopes to the south, against a south-facing wall, or next to a white wall. Also, let the water sit in the sun to warm up before watering the plants.
- Trench tomato transplants by laying them sideways and pinching off the lower leaves below the soil level. Have only the top cluster, or no more than one-third of the plant showing. This makes for a sturdier plant, and it gets warmer faster. Keep the ground moist for three or four days after transplanting.
- Apply a mulch to the soil using straw or newspaper to conserve moisture and cut down on weeds. Mulches reduce fruit cracking and blossom-end rot by keeping the soil evenly moist. Apply mulch when plants are eight to ten inches tall.
- Add stakes to the tomato cages if the plants outgrow the cages. You can also tie plants to the stakes. Weave stakes through the cages, and anchor them in place. Although staked or caged tomatoes produce fewer fruit, they are usually richer in vitamin C from having received more sunshine; plus tomatoes from staked or caged plants stay cleaner and are easier to pick.
- Eliminate the need for stakes or cages by piling straw almost level to the top of the plants when they are about a foot high. The branches lie on the straw as they grow, and the straw keeps the tomatoes clean. The mulch will also hold in moisture and keep down the weeds.
- Space tomato plants about three feet apart to help prevent fungal diseases.

- Get a jump-start on the tomato season by encasing the plants in plastic. Cut the bottom out of a black plastic garbage bag, and slip it over the tomato cage. Mound dirt around the bottom of the bag to hold it in place, and secure the top of the bag to the cage with clothespins. Leave the top open to let in air and sunlight (see *TRANSPLANTING IN COOL WEATHER, hints on*).
- Sink a clay pot or coffee can (punch holes in the bottom of the can) next to the plants; then fill it with water once a week. This ensures a steady supply of moisture to the roots and expedites fertilizing; just add diluted fertilizer to the water.
- Increase pollination by hitting the tomato stakes or cages or gently shaking the plants during the driest part of the day.
- Do not cut off any foliage during early stages of growth. Tests have shown that pruning or pinching hinders production. However, when the tomatoes get too tall for stakes or cages, cut them off at six feet or when they reach the tops of the cages. This will encourage more side shoots and bigger fruit production.
- Hand-pick hornworms by sprinkling the plants lightly during early morning. The hornworms will shake off the water, making them easier to see and remove. Drop them in a can of soapy water.
- Cut back on watering when fruits begin to ripen. This way, the sugars can develop, and the tomatoes will be sweeter.
- Sterilize the stakes and cages at the end of the season: Wipe off the soil and other residue, and then wipe the supports with a cloth dipped in a solution of one part chlorine bleach to nine parts water. Leave the bleach solution on for about ten minutes, and then rinse well and dry.

***TOOLS, GARDEN, hints on***

- Paint a band of bright-red or orange paint on the handles of your garden tools. Or tie colorful ribbons to the handles. This makes them easy to find if they are left in the garden.

- Paint marks on hoe and rake handles at twelve-, eighteen-, and twenty-four-inch intervals for quick measuring. Mark a trowel for proper bulb-planting depth while you're at it.
- Sharpen pruning blades and knives with an oiled honing stone. Slide it in one direction across the tools' beveled edges.
- Use a child's rake to remove dead leaves from a perennial bed.
- Wrap short strips of pipe insulation, carpet padding, or moleskin around the handles of tools to prevent blisters. Or slip sections of an old bicycle inner tube over the handles. The rubber will make the work easier on your hands.
- Place rubber crutch tips on the ends of tools, or use the rubber tips from old mops or brooms to make them easier on the hands.
- Remove the top and bottom from a large empty container (such as a 5-pound coffee can), and nail it to a fence post. Position the container about three feet from the ground, and use it to hold long-handled tools.
- Put an old mailbox in the garden, or attach it to the garden fence to hold small tools, gardening gloves, and other small items. Or use a gallon-size plastic bleach bottle by cutting away the upper side opposite the handle.
- Buy a child's red wagon at a garage sale, and use it to cart heavy items around the garden.
- Use an old golf bag to haul rakes and shovels around. Pockets on the bag can be used for small tools, gloves, and so on.

***TOOLS, GARDEN, substitutes for***

- Use old knives of various sizes for different chores: a butter knife for picking out seedlings for transplanting, a putty knife for separating transplanting blocks, a grapefruit knife for digging out weeds, and a bread knife for harvesting greens. Also, try using a sharp box cutter or large penknife for cutting flowers; it's easier to handle than a regular knife and makes a faster, cleaner cut.

- Use old forks. The tines of large forks get under weed roots better than most tools, and salad forks are great for aerating the soil of houseplants.
- Use an old ice-cream scoop to dig uniformly sized holes for transplanting, or use a melon-ball scoop to make small holes for tiny bulbs.
- Use a discarded colander or kitchen sieve as a substitute for a screen to sift soil covering indoor seeds or outdoor plants.
- Use an insert from a salad spinner for spreading grass seed or broadcasting wildflower seeds over a large area.
- Use an old screwdriver to pry up weeds.

***TOOLS, GARDEN, to prevent rusting of***

- Clean the metal parts of small tools after each use: use a stiff-bristle brush or emery cloth; then apply a protective coating of oil or petroleum jelly before putting them away.
- Clean and lubricate the tools in a mixture of sand and vegetable oil. Fill a pail with sand, pour in one quart of oil, and mix it together. Keep the container in the garage or tool shed; after you use the tools, push them into the oily sand a few times. If you prefer, simply use a pail of plain dry sand to store the tools.
- Rub the lawnmower with an oily cloth after each use.
- Spray lubricating oil, such as WD-40, on rakes, shovels, and pruning shears before storing.
- Consider purchasing small tools made of aluminum alloy or stainless steel, which never rust.
- Put a tool that is rusted beyond repair to good use. Cover it with water, and let it sit until the water turns brown. Dilute the solution with fresh water and then pour it around the plants to give them an iron-mineral boost (see *IRON FOR PLANTS, hints on*).

***TRANSPLANTING FROM PEAT POTS;*** see *PEAT POTS, TRANSPLANTING FROM, hints on*

***TRANSPLANTING IN COOL WEATHER, hints on***

- Plant vegetables in raised beds. Raised beds help spring soil drain excess moisture faster and keep plant roots five to ten degrees warmer because they're exposed to more sunlight.
- Lay down clear heavy plastic ten days before planting. Clear heavy plastic warms soil anywhere from six to fifteen degrees. (Check nurseries or commercial growers to see whether you can obtain their old plastic glazing from greenhouses.)
- Make plastic tunnels to protect the plants from the cold: drape heavy plastic over wire hoops or frames fashioned from wire coat hangers. Use boards, stones, two-by-fours, or long metal pipes to weigh down the edges of the plastic. Close the ends during cold nights.
- Cover the plants with heavy double row covers. Just put one cover on top of another, and close the ends. Or protect individual plants with hot caps (see *HOT CAPS, hints on*).

***TRANSPLANTING IN HOT WEATHER, hints on;*** see also *SEEDS, to sow in hot weather, hints on; SEEDLINGS IN THE GARDEN, to shade*

- Soak the soil thoroughly the day before transplanting, and transplant after the sun goes down when it's hot.
- Dig the hole for the new plant a little deeper than usual, and put some shredded newspaper in the bottom. Fill the hole with water, and let sit until absorbed. When all the water is gone, set the plant in place, cover the roots with dry soil, and then water very gently to settle the soil. The newspaper will retain more water in the area as it decomposes.
- Plant transplants in the shade of annual flowers or vegetables, or provide shade until the plants are established.
- Pour about two cups of seaweed emulsion around each transplant to encourage root growth. Or use it as a foliar spray. Kelp extract has natural growth hormones. (Use one tablespoon of emulsion per one gallon of water.)

***TREE SEEDS, to propagate***

- Gather the seeds from fruit trees when the seeds are ripe and ready to fall from the tree. Stick the seed just under the soil surface, and water it well. Let the soil dry out before the next watering. If you plant the seeds in a container, cover the container with a plastic bag in which you've poked a few holes. Place the container in a warm, bright location, and occasionally spray the topsoil with warm water to keep the humidity level high. Remove the plastic when the seed sprouts. For orange seeds, wash them in lukewarm water, and dry them before planting.
- Put the tree seeds in with a houseplant that likes a moist environment; then repot the seedlings when they have their first sets of true leaves.

***TREE STAKING, hint on***

- Secure the tree to the stake with a piece of an old bicycle inner tube, old thin hose, or canvas strapping.

***TREES, FRUIT, hints on***

- Wash fruit-tree foliage periodically to remove dust and honeydew secretions and to dislodge any pests. Clean foliage also encourages beneficial insects.
- Cultivate moss or ground cover under fruit trees to offer shelter and food for beneficial predators and to keep down the weeds. This will also prevent fruit-fly larvae from making contact with the soil and continuing their life cycle.
- Plant tansy under the trees, especially peach trees, to protect them from borers.
- Prevent gypsy moth caterpillars and female cankerworms from crawling up trees by using a barrier. Wrap a three- to four-inch strip of fabric or pantyhose tightly around the lower trunk, and then on that put a strip of plastic wrap or duct tape. Spread petroleum jelly, axle grease, or Tanglefoot on the plastic or duct tape, leaving a half-inch margin at the top and bottom of the band. Put the barrier in place before the fall and

spring seasons. When the trap loses its effectiveness, replace it, and add more sticky substance.

- Control aphids by putting up a sticky barrier as directed above. This will prevent ants from crawling up the tree to farm the aphids for their honeydew.
- Avoid wormy fruit by hanging traps in the trees just before the blossoms are about to open. Combine and thoroughly mix one cup of warm water, one-half cup of apple cider vinegar, one-quarter cup of sugar, and one tablespoon of molasses. Pour about one inch of the solution into small containers, such as yogurt or Styrofoam cups, and hang two or three in each tree. Empty and refill the traps as needed.
- Control spider mites by spraying with horticultural oil spray one to two weeks after the petals fall (see *HORTICULTURAL OIL SPRAY*).
- Work in some wood ashes under the trees, or spray the tree with diluted seaweed extract when blossoms appear. Fruit trees need potash to make fruit sugar.
- Provide frost protection during flowering by spraying the tree with seaweed emulsion. The kelp will prevent loss of fruit set during low temperatures.
- Pinch off surplus fruit when the fruit is pea size. Leave about two and a half inches (or the span of your hand) between each fruit.
- Use an old rake to prop up fruit-laden branches. Tuck the branch between the rake's tines, and cushion the branch with an old glove or a piece of folded fabric. Angle the rake handle into the soil so the branch can still sway in the breeze.
- Cover ripening fruit with small paper bags to protect it from bugs and birds. Cut a small hole in the bottom of each bag for ventilation and to let rainwater escape.
- Keep both hands free when picking fruit by wearing a large lightweight shoulder bag strapped across your body. Another way to keep both hands free is to loop a bag or pail through a belt or cord fastened around your waist.

***TREES, information on***

- Call your county forester or the Forestry Division of the county for recommended trees to plant in your area or advice on those you already have.

***TRELLISES, hints on***

- Erect a trellis toward the back of the garden and on the north side so it doesn't overshadow other plants.
- Train the climbing plant on both sides of the trellis.
- Suspend a three- or six-foot-high strip of chicken wire between two stakes. Or support the wire with stakes at three-foot intervals. Weave the stakes through the wire, and sink them one foot deep into the soil.
- Weave twine along the garden fence, and use it as a trellis. Or suspend heavy-duty nylon netting from the fence.

***TULIPS, hints on;*** see also *BULBS, hints on*

- Protect the tulips from squirrels, gophers, and mice by doing any of the following:
  1. Cover the bulbs with pea-size gravel or crushed oyster shells before covering them with soil.
  2. Lay a wire screen or hardware cloth over the bulbs, and remove it in early spring.
  3. Soak the bulbs in Ropel; then let them dry thoroughly before planting. (Ropel is available in garden-supply stores.)
- Refrigerate tulip bulbs for a minimum of six weeks before planting in mild-winter areas. Store the refrigerated bulbs in paper bags and away from ripening fruit. The fruit gives off ethylene gas, which is detrimental to bulbs. Or buy special cold-treated bulbs for southern planting. (Be sure not to freeze the bulbs.)
- Water bulbs after planting, then sparingly until leaves emerge, and then generously thereafter.

- Cut tulips for indoor arrangements just as they are starting to open and before sunup or after sundown. Be sure not to take any foliage. Stand the tulips up to their heads in lukewarm water, and leave them in a cool place for twelve hours. Add a few drops of vodka to the water to keep them standing straight. Tulips will last a week if conditioned this way.

### *TURNIPS, hints on*

- Plant turnips later in the season than normal. Late-planted turnips are more resistant to maggots than those planted early in the spring.
- Interplant turnips with peas, which are friends, but not with potatoes, which are not compatible with turnips.
- Harvest turnips when they are no more than three inches in diameter; otherwise they become woody and bitter. Loosen the soil around the base of the leaves, and then pull up the roots by the tops.

# V

***VEGETABLES, hints on***

- Pick a sunny location for a vegetable garden, not under trees or where water can collect. Leafy vegetables such as lettuce and spinach can be grown in partial shade, but fruiting vegetables such as squash and tomatoes need at least six hours of sunlight a day.
- Avoid growing vegetables in the same soil two years in a row (preferably three years), or in soil previously planted with anything from the same botanical family (e.g., don't follow potatoes with tomatoes, or squash with cucumbers). If the vegetables are kin, they are prone to the same pests and diseases. Also, alternate deeply rooted vegetables, such as carrots, with shallowly rooted ones, like lettuce.
- Avoid purchasing vegetable transplants that are already flowering. Buy transplants that are small and compact; they will adjust better when transplanted.
- Buy transplants singly rather than in a six-pack of the same kind. This way you can try an assortment of new varieties along with a plant or two of tried-and-true, reliable favorites. Another benefit of buying single transplants is in case of disease, pests, or drought; one variety may be more resistant than another, and you can also get early, midseason, and late varieties to stretch the harvest.
- Grow what does well in your area, what you enjoy eating, and only as much as you will eat, freeze, store, or give away.
- Plant high-priced vegetables or those not available in your local market. Most produce is seasonally priced, but gourmet or specialty items are not; these are never inexpensive. Certain other vegetables don't ship well and are not marketed by commercial growers.

- Plant vegetables in square grids or in raised beds. These methods allow closer planting and produce more vegetables per area of garden space. In closer planting, the crop leaves shade the soil, which blocks out the sun, conserves water, and cuts down on weeds.
- Make raised beds no more than three feet wide; that way, you can cultivate and harvest from either side of the bed.
- Stagger plantings, and make new sowings every ten days to two weeks, or plant early-, mid-, and late-season varieties to avoid a glut when everything matures at the same time.
- Plant beet, cucumber, and squash seeds directly in the garden. Beets have long taproots, and cucumber and squash roots are difficult to disentangle without disturbance.
- Provide one inch of water every week, keep the soil adequately moist at all times, and never let the plants dry out between watering. Quick, steady, uninterrupted growth is the formula for maximum yields and quality vegetables.
- Leave light row covers or other protection on carrots and onions all season. Uncover all other crops, such as beans or cabbage, once the generation of pests is past. Always remove coverings from fruiting vegetables when they are starting to flower so pollinating insects can fertilize the blossoms. It's also a good idea to remove row covers in hot climates.
- Interplant two, three, or more crops, and combine companion plants. Interplanting increases yields, improves soil quality, and reduces pest and disease problems.
- Start harvesting as soon as something is big enough to eat. This will give you the earliest, most tender vegetables, make room for the remaining ones to branch out, and stretch the harvest.
- Plant vegetables that don't require frequent harvesting if you're a weekend gardener: cabbages, collards, eggplants, kales, lettuces, peppers, pumpkins, root crops, tomatoes, and winter crops.

- Pinch off tops of vine-growing species and any blossoms or fuzzy ends of vegetables or fruits late in the season (six weeks before the prospect of frost) so the plants can concentrate on fruit production.
- Pick most vegetables late in the day, when carbohydrates and proteins are at their highest concentrations. Pick crisp vegetables like lettuce, broccoli, and cucumber in the morning, which is when they are at their best.
- Hose off root vegetables in the garden in a dish drainer, mesh laundry basket, or plastic colander. This way, any loose soil can be returned to the earth. Drain off excess water from the containers before bringing them inside.
- Turn under end-of-season crops right away if they're not destined for the compost heap. Chop them up with a shovel first, dig them into soil about six inches deep, then chop them up a bit when they are in the soil. Give the crop residue two weeks to start decomposing before doing any further planting.
- Clean vegetables by soaking them for a few minutes in water with a little table salt added. This will kill any bugs lurking in the produce. Salt makes lettuce and other salad greens flabby, so use vinegar or lemon juice instead.

### *VEGETABLES, EASY-CARE*

- The following vegetables are easy to care for: beets, carrots, kohlrabies, lettuces, mustard greens, radishes, runner beans, sprouting broccolis, Swiss chards, or zucchinis.

***WATER CONSERVATION, information on***

- Plant native plants and smaller plants rather than larger ones to conserve water.
- Choose drought-tolerant plants in drought areas. Your local County Extension Office or Cooperative Extension Office is an excellent resource for native plants in your region. Or your local botanical garden or garden center may have a list of water-thrifty plants suited to your area.
- Download a copy of *California Friendly*, a book on water-conscious gardening that is available for free through the Metropolitan Water District of Southern California's Be Water Wise program. Go to BeWaterWise.com, and click on "ToolKit." From there, you can get a PDF version of the book. (See also Internet Resources for additional water-conservation downloads.)
- Water less frequently, but water deeply until the water reaches the plants' roots.
- Reduce fertilizer use to the lowest possible level.
- Consider drip irrigation with flexible plastic tubing and emitters that run to each plant. This works well for smaller gardens that contain ornamental trees, shrubs, perennials, and ornamental grasses, and in vegetable beds. You can plan your own system or buy prepackaged vegetable-garden irrigation kits. Another benefit is that weed seeds don't get watered, and fewer weeds grow. Check the drip system periodically for leaks and clogs.
- Use soil-soaker hoses for small areas and vegetable gardens. Cover them lightly with bark mulch, hay, or straw in a semi-permanent place, if desired. Soaker hoses water the plants' roots without wetting the leaves,

using 25 to 40 percent less water than an overhead sprinkler, and reduce the likelihood of disease.

- Use olla or pitcher irrigation for vegetables by sinking unglazed clay pots (without holes) in the ground with their tops about an inch above the surrounding soil. Fill them with water—one to one-and-a-half gallons—and then cover them with lids. The buried porous pots will slowly irrigate an area three to four feet in diameter. (Substitute a regular unglazed clay pot for an olla pot by sealing the hole with a rubber stopper, a wine cork, or a piece of tile held in place with waterproof glue. Make a lid from wood, floor tile, or other material, or use a plant saucer or old ceramic plate. If you want to collect rainwater, drill a small hole in the lid.)
- Create container irrigation for new trees and shrubs by sinking plastic gallon milk cartons or kitty-litter containers with one or two small holes made in the bottom and filling them with water. The water will seep into the soil.
- Use lightly used water from sinks, showers, baths, and the washing machine for trees, bushes, shrubs, large annuals, and perennials (or smaller plants growing closely together). Use the water on food crops with the edible portion above ground, such as tomatoes, corn, peppers, and squash, so the water doesn't touch the vegetables. Avoid using it on seedlings and on crops sensitive to salts: avocados, citruses, and various herbs. And because gray water is alkaline, avoid using it on acid-loving plants. (Use plant-friendly, gray water–compatible liquid soaps, which are very low in salts and free of boron and bleach; and use the gray water right away; don't let it sit around).
- Place a rain barrel (or a large heavy-duty plastic barrel or garbage can with an attached hose connector) under the rain gutter to catch rainwater. Cover the top of the barrel and downspout with aluminum or fiberglass window screening or hardware cloth to keep out debris and mosquitoes. (Do not use a rain barrel if the roof is made of asphalt or treated-wood shingles.) If necessary, keep the water in the rain barrel fresh by adding some crystals of potassium permanganate at regular intervals. This

has no adverse effect on plants or seedlings. Another method is to use unscented chlorine bleach (1 ounce added to 55 gallons of water; then left 24 hours before using to allow the chlorine to dissipate). If mosquitoes are a problem, use BT (*Bacillus thuringiensis*), which kills mosquito larvae and does not cause harm to pets, fish, or humans. Or, empty out the barrel once a week, and clean it on a regular basis.

- Add polymer granules to the soil to cut watering by 50 percent. The granules can absorb up to ninety times their weight in water, are nontoxic, last for several seasons, and eventually disintegrate into the soil. Use one-half teaspoon per planting hole for bedding plants and vegetables.
- Plant vegetables within shallow depressions: sunken beds with excavated soil piled into ridges around each square. Or provide irrigation ditches: shallow trenches along one or two sides of each row. This will reduce water needs and prevent runoff.
- Apply a mulch whenever possible (around plants and in bare soil). This will cut down on watering and conserve our natural resource (see *MULCH MATERIAL, hints on*).

***WATERING A YOUNG TREE, SHRUB, OR WATER-LOVING PLANT, hints on***

- Poke a tiny hole in the bottom or near the base of a plastic gallon-size juice or bleach bottle; then put it next to the tree, shrub, or plant. Or partially bury the bottle upright between two plants. Fill the jug with water once a week or as needed; the water will drip out slowly and seep down to the roots. This watering method is particularly useful on a slope, where water tends to run off before it can be absorbed.
- Punch a hole in the bottle cap of a plastic gallon-size bleach bottle or other container, and lay the bottle on its side, slightly tilted so the water can drip out slowly. The water collects heat from the sun, which is a bonus for heat-loving plants and for cucumbers, which like warm water.
- Drive PVC pipes into the ground near water-loving plants such as tomatoes, peppers, cucumbers, and melons, and then pour water into the pipes. The water will get below the root level.

- Sink a clay flowerpot in the ground with the rim an inch above soil level; then stuff a wad of cloth tightly into the drainage hole, and fill the pot with water. This releases the water in a slow dribble.
- Line the sides of terra-cotta pots with heavy plastic or polythene Bubble Wrap to reduce water loss. Keep the draining holes clear.

***WATERING CAN, hints on***

- Choose a watering can that can fit under the kitchen faucet as well as the outside faucet, and make sure the can is light enough to carry when filled with water. Also, look for one with a fine rose for watering delicate seedlings.
- Use an empty plastic gallon-size milk or laundry detergent jug. Wash the container thoroughly, then make holes in the cap with a heated nail or needle.

***WATERING WAND, to make***

- Fasten a four-foot length of an old broom or mop handle along the end of the garden hose. Use duct tape or other strong tape to attach the wood to the hose, and then use the wand to direct the spray of water around the base of plants.

***WATERING, hints on***

- Water in the morning, preferably around sunrise if possible, when winds are at a minimum. Doing so will give the plants a chance to dry, since fungal diseases thrive on wet foliage. It will also prevent water from being lost to midday evaporation.
- Fasten a piece of burlap or an old cotton sock over the hose end so the ground will not be washed away beneath the stream of water. Or place a board or rock under the hose if leaving the hose to slowly water a specific area.
- Soak established trees several feet beyond the canopy of the leaves; soak young trees just up to the canopy.

- Irrigate plants at ground level, with the exception of fuchsias, tuberous begonias, azaleas, and others that benefit from daily overhead sprinkling throughout the growing season. Make sure, however, that the roots also receive adequate water.
- Keep an eye on chrysanthemums as an indication of sufficient water; they are the first plants to wilt when water is scarce.
- Make sure that most plants receive one inch of water a week, with the exception of cacti and other drought-resistant plants.

***WEED KILLER, to make***

- Add two tablespoons of mild liquid dish soap to one gallon of distilled white vinegar. Avoid spraying the solution directly on the soil.

***WEED-KILLER SPRAY, hints on applying***

- Apply weed killer on a dry, calm day, preferably during late-morning hours.
- Apply granular weed killer on wet, watered areas.
- Avoid spraying adjoining vegetation. To protect nearby plants, cover them with plastic sheeting. To target selected weeds, sprinkle the herbicide through an empty toilet-paper cylinder, or spray the solution through a clean empty food can or a plastic bottle with the bottom removed. Wait a second or two for the spray to settle; then lift the cylinder, and move to the next weed.

***WEEDS AND GRASS, to control;*** see also *MULCH, hints on*

- Eat the weeds you can identify and know are safe. Weeds are more nutritious than cultivated greens and can be eaten raw or cooked. Purslane, for instance, is said to contain more omega-3 fatty acids than any other leafy vegetable. The following are a few common safe-to-eat weeds: chickweed, red clover, curly dock, dandelion, lamb's-quarters, mallow, mustard, plantain, purslane, and wild amaranth.
- Lay old magazines along the edges of flower and vegetable beds, fences, and other areas to keep them free from encroaching weeds and grass.

Cover the magazines with an inch or more of soil to hide them. Doing this will also keep the beds' borders neat and cut down on the need for trimming.

- Prevent grass from invading flower beds by burying steel or plastic edging six to eight inches deep between the lawn and the beds.
- Put down black plastic, old carpeting, cardboard, or a three- to six-inch-thick layer of newspapers in garden paths and walkways to discourage weeds. Cover with soil or bark.
- Water newly prepared beds every couple of days for two weeks before sowing seeds or setting out transplants. This will allow as many weed seedlings as possible to germinate before planting. Hoe the weeds off as they appear.
- Stir the top quarter-inch of soil every week to expose the annual weeds. (This dust-mulch cultivation only applies to unmulched areas.) Or avoid having too much bare soil in garden beds by placing plants closer together, as found in nature, and leaving any fallen leaves in place.
- Use corn-gluten meal on the lawn and in the perennial border in the early spring, but not in the areas for growing plants from seed. A byproduct of processing corn, corn-gluten meal is a nonhazardous preemergent herbicide that prevents seeds from sprouting and an alternative to synthetic preemergent chemicals. It is also a quality nitrogen source without any of the environmental side effects of a typical weed and feed.
- Take a bucket or tarp with you when weeding; it makes garden cleanup easier.

***WEEDS AND GRASS IN PAVING, PATIO CRACKS, AND WALKWAYS, to remove;*** see *GRASS AND WEEDS IN PAVING, PATIO CRACKS, AND WALKWAYS, to remove*

***WEEDS AND GRASS IN SOIL, to remove;*** see *GRASS AND WEEDS IN UNPLANTED SOIL, to remove*

***WET OR POORLY DRAINED AREA;*** see *SOIL, POORLY DRAINED, hints on*

***WHITEFLIES, to control***

- Hose adult whiteflies off plants, hitting both sides of the leaves, and repeat every few days.
- Apply insecticidal soap, pepper spray, or place sticky insect traps near the plants (see *INSECTICIDAL SOAP; PEPPER SPRAY; STICKY INSECT TRAPS, to make*).

***WILDFLOWERS, hints on***

- Buy wildflower mixes that are local to your region.
- Plant wildflowers in fall and spring, which are the best times to sow them.
- Improve wildflower germination rates in cold-climate areas by putting seeds in the freezer for two days, and then thawing them for one day. Repeat the procedure a few more times.
- Add sand to the seeding mix to ensure even distribution.

***WILLOWS, hints on***

- Plant willows in a moist or wet area; otherwise they will seek out sewer lines and other sources of water.
- Start a new tree by cutting a two-inch-long branch about one-half inch across; then plant it in moist soil.

***WINDOW BOXES, hints on***

- Line the bases of window boxes with several layers of newspaper before adding the soil. The paper prevents the soil from drying out too quickly after each watering. Make slits in the newspaper for water to escape.
- Prevent plants from drying out too quickly in hot areas by using window boxes that are eight to twelve inches deep.
- Place a layer of gravel on the tops of the boxes to prevent the soil from splattering when it rains.

- Use rigid plastic liners that fit the boxes. This makes replanting easier; just lift the liner from the box, and replace it. Punch out the perforated drainage holes or drill holes.

***WINDOW BOXES, plant choices for;*** see *FLOWERS, best for window boxes*

***WIREWORMS, to control***

- Bury peeled and sliced potatoes one inch in the ground with a stick through them that protrudes out of the soil. Pull out and replace the traps every couple of days.

***WOOD CHIPS, hint on***

- Fertilize the area first with a high-nitrogen fertilizer before using wood chips. As the chips decay, nitrogen is temporarily depleted from the soil. Wood chips work best, though, for garden pathways where they will not be disturbed by digging.

***WOODCHUCKS, to control***

- Spray plants with a deterrent solution of garlic and hot pepper (see *RABBITS, to repel*).
- Put row covers or other protection over the vegetables.
- Sprinkle cayenne pepper in and around the woodchucks' burrows and the crops they favor. (Buy the cayenne pepper in bulk.)

# Y

***YELLOW JACKETS, to get rid of***

- Boil together one-quarter cup of sugar and three-quarters of a cup of water. Cool, and then put the liquid in a small container with a lid. Make a small hole in the lid, and hang the container in a tree. The wasps will crawl in the hole and be unable to get out. Dispose of the wasp trap by placing it in a plastic bag and sealing the bag carefully before putting it in the trash.
- Put two inches of white vinegar in a long-neck bottle, and leave it out to attract wasps.
- Cover a soil-level nest with a clear-plastic or glass bowl, and leave it in place until you are certain all the yellow jackets have succumbed. Cover the nest at dusk, when the yellow jackets have turned in for the night.
- Set out traps only in places frequented by people or pets or if a family member is allergic to insect stings.

# Z

**_ZINNIAS, hints on_**

- Water zinnias at soil level, never overhead, or you'll ruin the foliage and encourage mildew.
- Provide optimum conditions for zinnias to avoid mildew. Make sure they are in a sunny location with good air circulation, and thin out extra stems if necessary to ensure air movement.
- Add a few drops of chlorine bleach to the water for a cut-flower arrangement. Zinnias can discolor the water pretty fast.

**_ZUCCHINI, hints on;_** see also *SQUASH, hints on*

- Be conservative when planting zucchini. One or two plants should be more than enough for a family.
- Pick the zucchini at least every second day and when they are no more than four to five inches long. Otherwise the vines may halt production.

# Indoor Gardening

***ACID-LOVING PLANTS, hint on***

- Water gardenias and other acid-loving plants with diluted leftover coffee or weak tea once a month, or with a solution of a few drops of lemon juice or white vinegar in one quart of warm water.

***AFRICAN VIOLETS AND GLOXINIA, hints on***

- Use a potting soil specifically designed for African violets or a mixture of half sphagnum moss and half perlite.
- Provide African violets and gloxinias eight hours or more of good light a day in a window providing indirect diffused light and free from cold or warm drafts.
- Keep African violets or gloxinias in small pots about two-and-three-quarters to three inches in diameter. These plants need to be root bound to bloom.
- Set plant containers over trays of pebbles filled with water. The containers should be in saucers so the pots don't stand in water.
- Water the plants from the bottom, and use a soft brush to clean the leaves. Once a month, water from the top, being careful to avoid wetting the crown, the point on the stem just above the soil where the leaves begin to branch off. The crown tends to rot if it remains wet. (A turkey baster works well when watering from the top.)

- Use warm water when watering the plants, preferably rainwater or distilled water. Doing so will reduce the amount of salt buildup. Also, leaves that are splashed with cold water may develop yellow spots.

***BENEFICIAL HOUSEPLANTS;*** see *PLANTS TO PURIFY THE AIR*

### *BROKEN STEMS, to fix*

- Make a splint with a toothpick or matchstick, and hold it in against the broken stem area with tape.

### *BROWN ENDS, to trim*

- Trim off the brown ends at the angle in which the leaves grow, in an inverted V, not straight across, and leave about one-half inch area between brown and green, or else the leaf will bruise and brown further. Often tip burn or scorched edges are due to lack of humidity.

### *BULBS, FORCING, hints on*

- Bury bulbs in a pot of vermiculite, sand, or gravel with the tops sticking out. Water the mixture until it's damp, and then keep the pot in a dry, cool place. Do not let the mixture dry out. When shoots appear, place the pot in the sun or a sunny window.

### *CACTI, hints on*

- Feed cacti once a season with quarter-strength plant food or a little bone meal scratched into the soil.
- Repot cacti when dry and during spring, when growth is active. Let the soil dry out, repot into dry soil, and don't water for one or two weeks.
- Protect your hands when repotting cacti by using a pair of tongs or heavy oven mints.

***CARE OF INDOOR PLANTS, hints on;*** see also *FERTILIZING, hints on*

- Dust the plants with a feather duster, or blow the dust off with a hair dryer set on cool. Mist tropical plants afterward.
- Occasionally spray or mist nonhairy plants in the bathtub, and then let them drip until dry enough to return to their customary places. A faster way to water the plants is to put them under the shower at the lowest pressure. Cover the soil with a piece of foil or plastic if too much water hits the soil directly.
- Turn the plants on a regular basis, such as every week or whenever you water them. This will ensure that all parts will get their ration of light and grow symmetrically. (Use a lazy Susan as a base for a large plant so it can be turned easily.)
- Aerate the soil with a fork occasionally.
- Clean the leaves of large plants by wiping them gently with a damp soft cloth or a soft cloth that has been dipped in a solution of one tablespoon of white vinegar per quart of cool water.
- Wash or spray dusty or dirty leaves with a solution of one teaspoon of mild dish soap per quart of lukewarm water; then rinse or spray afterward with clean water. Do not wash African violets or other plants with hairy leaves.
- Apply mulch to conserve water, but keep it an inch or so away from the plant stems.
- Freshen the soil occasionally by removing some of the old soil on top and replacing it with fresh soil.

***CATS, to keep away from plants***

- Crush rue leaves, and put them on the soil surface, or sprinkle powder made of cayenne pepper and black pepper on the soil to keep cats away from plants.
- Cover the top of a container with netting, screening, or strips of tape to prevent the cat from using the plant container as a litter box.

- Grind up a few alfalfa tablets (not alfalfa sprouts), and mix them with the cat's food. This might prevent the cat from chewing the plants.
- Grow a pot of wheatgrass for the cat, and make new plantings every few weeks. (Obtain the wheat berries at a health-food store.) Or buy a small flat of wheatgrass at a farmers' market.

***CHRYSANTHEMUMS, hint on***

- Cut back chrysanthemum plants after flowering, leaving a few green leaves on the stems. Take the plants out of their pots, and divide them into separate sections. Plant the sections in the garden. When new growth sprouts from the base, remove the old stems.

***CUTTINGS, PLANT, hints on***

- Propagate creeping Charlie, coleus, ivy, or pothos by taking three- to four-inch cuttings one-quarter inch below a leaf node and putting them in a jar two-thirds filled with water. Strip off any leaves, except for a few at the top. Cover the jar with plastic wrap or aluminum foil, poke small holes in the cover, and stick the cuttings through the holes. Keep in a warm, light place out of direct sunlight. Remove the cover every few days, and stir the water to incorporate oxygen. Plant the cuttings in soil while the roots are still small so they can adjust easier.
- Pothos cuttings can grow in a vase of water indefinitely. However, do not add fertilizer to the water, as it could burn the roots.
- Save used Styrofoam coffee cups and soft-drink containers (tops removed) to root cuttings (wash the cups and containers thoroughly).

***EASTER LILY, POTTED, hint on***

- Continue to water the Easter lily occasionally after the flowers fade and until the leaves are brown and dry. Then cut a few inches off the top of the stem, and plant it outdoors in a sunny, well-drained location.

**EASY-CARE PLANTS**

- The following plants are easy to care for: aloe vera, aspidistra (a.k.a. cast-iron plant), cactus, dracaena, English or Swedish ivy, jade plant, kalanchoe, philodendron, rubber plant, schefflera, snake plant, and spider plant.

**FERNS, hints on**

- Provide ferns with shade and steady moisture; otherwise they will be short lived. Northern light in a cool room or in a bathroom that provides humidity suits them fine.
- Water ferns occasionally with leftover tea. Or mix soggy tea leaves into the potting soil. Ferns thrive on it.
- Keep ferns on a low-fertilizer diet. Ferns do not require as much feeding as most plants.
- Avoid using insecticidal soap on ferns. It could damage the foliage.

**FERTILIZING, hints on**

- Feed plants during rapid growth (summer) with a water-soluble fertilizer, such as fish emulsion or seaweed extract, every two to three weeks, or with a diluted amount every time you water. During slow growth periods, feed them every four to six weeks.
- Give plants a little praise or word of encouragement. Talking to your plants really does work, according to research.
- Water with leftover flat club soda; it contains beneficial minerals.
- Use the water from boiled eggs for a boost of calcium and potassium. Let it cool first.
- Mix three or four eggshells into one quart of water, shake it well, and let it sit overnight. Use the water on the plants. Egg whites and the matrix of the shell contain protein, which produces nitrogen.
- Use unsalted water from cooking or steaming vegetables as a fertilizer. Cool before applying.

- Add one drop of ammonia to one quart of water to improve the color of foliage and increase its growth, especially for ivy. Apply once a month.
- Try applying liquid seaweed extract for ailing plants. Liquid seaweed extract contains more than sixty minerals and can be applied to the soil to help remedy possible deficiencies, or it can be used as a foliar spray.
- Treat an ailing plant with a few drops of castor oil dribbled on the soil, followed by a thorough watering. The castor oil will also keep a healthy plant greener and improve blossoms.
- Never feed a sick plant or any plant in trouble. You will only hurt it.

### *HANGING PLANTS, hint on*

- Train trailing plants to cover the chains of the hanging basket by winding the stems gently around the chains or tying them loosely with twist ties.

### *HANGING PLANTS, to water*

- Before planting, line a hanging pot with plastic (or an opened-up plastic bag). Poke holes in the bag for drainage before putting the soil in. The lining helps the pot hold moisture longer. As an extra precaution against drips, place a coffee filter or a used dryer sheet in the bottom of the pot, or buy a saucer that clips onto the bottom of the container.
- Slip an old shower cap over the bottom of the container before watering the plant; then later carefully remove the cap with the dirt and drips.
- Put a few ice cubes into the pot, and let them melt.
- Buy a water wand from the hardware store. You squeeze the bulb, and the water shoots into a high, curved tube. Or use a large sports water bottle. The bent plastic straw makes reaching the containers easier.
- Soak a dried-out plant in a gallon of lukewarm water to which one or two drops of mild liquid dish soap have been added. Let the pot soak until the top of the soil feels moist. Adding soap to the water breaks the surface tension and helps rewet the potting mix.

***HUMIDITY FOR PLANTS, hints on***

- Mist most plants frequently, or group several plants together to increase humidity.
- Set pots in a waterproof tray containing pebbles and water to increase humidity. Spread a one-inch layer of pebbles in the tray, and then add a half-inch of water. Refill the pebble trays with water when necessary.

***KALANCHOE, hints on***

- Buy a kalanchoe plant in bud stage to have it last longer. The small bright flower clusters will bloom indoors for up to three months.
- Keep kalanchoes in indirect light. Direct sun can turn the leaves red and diminish the size of new growth.

***LOW-LIGHT-REQUIREMENT PLANTS***

- The following plants need low levels of light: African violet, aspidistra (a.k.a. cast-iron plant), Boston fern, Chinese evergreen, dragon tree, grape ivy, kangaroo vine, peace lily (a.k.a. spathe flower), philodendron, or spider plant.

***MOISTURE-LOVING PLANTS, hints on***

- Apply mulch to moisture-loving plants, but keep it an inch or so away from the plants' stems.
- Put one pot inside a larger pot, and insert sphagnum moss or cocopeat between the two pots. Keep the moss moist, and the potting soil will stay moist longer. Make sure both pots have drainage holes. This will also insulate patio plants from the heat.

***PESTS AND DISEASE, to control***

- Wipe flat leaves with a wet cloth to remove most insects. Or give the plants a firm jet of water to wash the pests off.
- Wash the plants with one-half teaspoon of mild liquid dish soap to one quart of water. Spray the tops and bottoms of leaves. Or if the

plant is small, invert it in a container of soapy water (hold the soil ball securely or cover it with plastic), and gently swish the leaves back and forth.

- Spot-spray a leaf before applying any pest-control spray. Then wait a day, and check for any damage or tip burn.
- To control aphids, squash them with your fingers, or run the plants under water to wash the bugs off.
- To control aphids and whiteflies, combine two teaspoons of mild liquid dish soap, such as Ivory or Dr. Bronner's Castile, and one quart of lukewarm water. Pour this into a spray bottle, and spray on plants as needed.
- To control fungus, spray the plants with a mixture of one-half teaspoon of baking soda combined with one quart of water.
- To control mealybugs or soft scale, pick off scale with a pair of tweezers. Or dip a cotton swab or cotton ball in a mixture of one part water to two parts rubbing alcohol (or vodka), and touch it directly to the mealybugs or soft scale. Repeat once a week until results are obtained. After each treatment, set the plants in the sink, and wash them gently with a spray of lukewarm water to remove any newly hatched insects.
- To control sooty mold, remove the pests that are causing the honeydew secretions and the mold (aphids, mealybugs, scale, and whiteflies) with a firm jet of water. Prune the area to increase light, and keep it dry. Sooty mold can also be a sign of high humidity and poor air circulation.
- To control spider mites, mist plants frequently. Spider mites thrive in a dry environment.

***PLANT POTS, hints on***

- Soak clay pots in clean water for a few hours before using them for planting; otherwise they will draw water out of the soil.
- Use clay pots for plants that like dry conditions and plastic pots for thirstier plants. Terra-cotta and clay keep plants cool but also absorb

moisture. Plastic pots heat up and cool quickly, minimize root damage in hot weather, and decrease water evaporation. Dark-colored pots absorb more heat; light-colored pots use less water.

***PLANT POTS, materials to cover the drainage hole in***

- Use cracked walnut shells, pottery shards, small stones, pebbles, Styrofoam packing pellets or broken-up packing material, activated charcoal, old sponges cut into squares, crushed rinsed eggshells, several layers of nylon net (or pantyhose, or mosquito netting), coffee filters, or used dryer sheets.
- Choose Styrofoam pellets, netting, or wire mesh to cover drainage holes in balcony or terrace containers. These materials will not add to the weight of the pots.
- Line the bottoms of containers without drainage holes with one or more inches of gravel or Styrofoam pellets (or broken-up Styrofoam). Use more for taller containers.

***PLANT POTS, to sterilize***

- Rinse pots to remove any soil; then scrub them with baking soda to remove any caked-on residue. Put them in the dishwasher with the clay pots on the bottom and the plastic pots on top, and then run the dishwasher through its complete cycle.
- Immerse pots in boiling water, or heat them briefly in a preheated 180°F oven. Then let them cool.
- Wash pots in hot soapy water, and then soak them overnight in a solution of one cup of chlorine bleach to nine cups of water. Rinse off the pots the next day and let them air dry.

***PLANTS TO PURIFY THE AIR***

- Plants that purify the air will absorb and neutralize air pollutants, such as toxins, benzene, and formaldehyde. Plants that purify the air include areca palm, Australian sword fern, Boston fern, dracaena, dumb cane, dwarf date palm, English ivy, golden peace lily, lady palm, pothos, reed palm, rubber plant, schefflera, snake plant, spider plant, and weeping fig.

***POINSETTIAS, hints on***

- Give poinsettia plants natural light but no direct sun.
- Punch a hole in the florist's foil to let excess water escape when you water the plant. Keep it moist at all times but not soggy.
- Clip the yellow flowers from the centers to have the blooms last up to a month longer.

***PUTTING PLANTS OUTDOORS, hints on***

- Give plants a treat, and put them outdoors when it rains. Avoid strong sunlight, and don't leave them outdoors for too long, or overnight; the change of climate may cause shock.
- Condition plants gradually if putting them outside permanently. Introduce them to outside conditions a few hours at a time, and then gradually increase the time they are outside. The reverse holds true for bringing outdoor plants indoors.

***REPOTTING PLANTS, hints on***

- Put down lots of old newspapers or a large, opened-up garbage bag to repot plants on; then scoop up the used soil and dispose of it.
- Prune the roots of container-grown plants in the spring or during active growth. Try this if you want to avoid having to repot to a larger container, but make sure the foliage and roots balance: if you prune one-quarter of the roots, remove one-quarter of the top growth.
- Wait two weeks or more before fertilizing a newly repotted plant.
- Repot cacti and succulents when dry. Let the soil dry out thoroughly, and repot the plants into dry soil; then don't water them for one or two weeks. Their roots are easily damaged.

***SOIL, hints on***

- Use soil mix or potting mix specifically formulated for container plants.
- Use lime-free soil mix to grow acid-loving plants, such as azaleas, rhododendrons, and gardenias.

***TROPICAL AND SUBTROPICAL PLANTS, Care of***

- Boil and cool tap water before watering tropical and subtropical plants to be on the safe side. A high chlorine content or too much lime in the water will poison a number of plants.

***VACATION CARE FOR PLANTS, hints on***

- Set houseplants in a bathtub containing one or two inches of water when you are leaving on vacation. Put the plants on old towels that are folded once or twice, or on thickly folded newspapers. The plants will absorb moisture as needed. Leave the bathroom light on if you have no exterior window.
- Keep plants healthy for up to one month by watering them well and then enclosing them completely but loosely in clear plastic bags, such as those that come from the dry cleaner. Tie the bags securely at the tops and bottoms. Place the plants in northern light, or where they will have light but no direct sun. When you return, untie the bags, and let the plants adjust to room air for a day before completely removing the covering.
- Bury one end of a heavy cotton cord or nylon clothesline just below the surface of the soil, and put the other end in a tall container of water. The cord or clothesline will wick up the water as needed.

***VEGETABLES SUITABLE FOR GROWING IN CONTAINERS***

- The following vegetables will grow in containers: bell peppers, chile peppers, eggplants, French beans (a.k.a. haricot verts), Hestia dwarf runner beans, herbs, lettuces, radishes, summer squashes, and tomatoes (patio or bush varieties).

***WATER-CONSERVING PLANTS***

- Choose a plant with small or needle-like leaves or a succulent such as agave. Other low-water plants are hens and chicks, orchids, ponytail palms, pothos, sago palms, snake plants, spider plants, and Tuscan blue rosemary plants. These plants are also low-maintenance.

***WATERING PLANTS, hints on***

- Water houseplants with room-temperature water, which is much better for plants than cold water since it can be more readily absorbed and speeds growth. Cold water shocks plants and actually slows growth.
- Water when the soil is dry below one inch from the top, and apply water until it runs out of the drain hole.
- Use a meat baster to siphon water out of the saucer if the plant is too heavy to move and the water has been sitting too long.
- Water a bone-dry plant by placing the pot in a container of water and letting it soak from the bottom up until the top feels moist. Also, give the plant a warm shower if the leaves are wilted. Omit the shower for African violets and other plants with hairy leaves.
- Don't water plants with softened water. Softened water has the calcium, magnesium, and iron replaced with sodium ions. Use water from an outside faucet or from collected rainwater. Or add one-half teaspoon of gypsum (calcium sulfate) to one gallon of softened water to provide the calcium needed to change it back to the unsoftened form.
- Let heavily chlorinated water sit overnight in an uncovered container. Exposure to air will dissipate the chlorine. Heavily chlorinated water can damage delicate plants, such as African violets, dracaenas, spider plants, and so on. To be on the safe side, boil the water, and let it cool overnight before watering chlorine-sensitive plants.

## Internet Resources

***BEES, BUTTERFLIES, AND POLLINATORS***

**The American Horticultural Society** has various articles on planting for pollinators:

ahsgardening.org/Gardening Resources/ Planting for Pollinators

**The National Wildlife Federation** has a PDF download on attracting butterflies to the garden, and a list of butterfly and moth species in your area and what host plants their caterpillars use:

nwf.org/backyard/Butterflies

**The Permaculture Research Institute** has a comprehensive list of beneficial insects and the specific plants they are attracted to:

permaculturenews.org/2014/Plants-Attract-Beneficial-Insects

**The Pollinator Partnership's** free pollinator- friendly planting guides are tailored to your specific area of the United States:

pollinator.org/Planting Guides

**The Xerces Society for Invertebrate Conservation** has various articles on attracting native pollinators to the garden:

xerces.org>Fact-Sheets

***BIRDS***

**The National Audubon Society** has a database of the best plants for birds in your area, as well as local resources and links to more information:

audubon.org/native-plants

**The US Environmental Protection Agency** has three PDF downloads that describe how to attract and feed different species of birds:

publications.usa.gov>Animals>Wildlife

***COMPOSTING***

**Compost-info-guide** has tips, articles, compost bin reviews, and a compost FAQ:

compost-info-guide.com

**The US Environmental Protection Agency** has fact sheets on composting and vermicomposting:

epa.gov/Composting

**The US Environmental Protection Agency** also has a PDF download "Backyard Composting: It's Only Natural":

publications.usa.gov>backyard composting

**The US Department of Agriculture** has links to various websites featuring composting:

nal.usda.gov/Compost-and-composting

**The USDA Natural Resources & Conservation Service** has a Backyard Conservation Tip Sheet with links to various websites featuring composting and vermicomposting:

nrcs.usda.gov/Composting

**The Cornell Waste Management Institute** provides access to a variety of composting educational materials and programs developed at Cornell University:

compost.css.cornell.edu/Composting Fact Sheets

## *GARDEN CATALOGES*

**Cyndi's Catalog of Garden Catalogs** lists more than two thousand mail-order garden catalogues in the United States and Canada:

gardenlist.com.

**Seeds of Diversity** lists the mail-order garden catalogues offered by Canadian companies selling open-pollinated, GMO-free fruit and vegetable seeds.

seeds.ca/Canadian Seed Catalog Index

## *GARDENING ADVICE*

**The USDA Cooperative Extension System** offers location-specific advice on various topics. Most states have gardening hotlines and offer information fact sheets, soil testing, and pest identification for free or a nominal cost:

nifa.usda.gov/Cooperative-Extension-System

**The USDA & NIFA Cooperative Extension** sponsors a network of knowledgeable experts from American land-grant universities that answer questions on garden-related topics:

extension.org/organic production

extension.org/mastergardener

*GARDENING COMPANIES*

**The Garden Watchdog** is a free directory that ranks more than seven thousand mail-order gardening companies on quality, price, and service, based on feedback from customers:

davesgarden.com

*GARDENING INFORMATION*

**Cornell University's Horticulture Department** has gardening guidance, and flower and vegetable growing guides:

gardening.cals.cornell.edu/Home

**The National Gardening Association** has articles on vegetables, flowers, herbs, and much more:

garden.org/Plants/Database

**The Royal Horticultural Society** has blogs, forums, articles, and a huge database of plants:

rhs.org.uk

**The United States Botanic Garden** has tips, articles, gardening fact sheets, native-plant recommendations, and a plant hotline:

usbg.gov/Grow/Gardening Fact Sheets

**The US Environmental Protection Agency** has a publication and PDF download "Greenscaping: The Easy Way to a Greener, Healthier Yard":

publications.usa.gov>Going Green> Recycling.

**The USDA National Agricultural Library** has information on flowers, produce, and landscaping:

nal.usda.gov/home gardening

*HERBS*

**The Herb Expert** offers comprehensive advice on growing and cooking with herbs:

herbexpert.co.uk

**The Herb Society of America** has fact sheets, guides, recipes featuring herbs, plus information on sustainable gardening practices:

herbsociety.org

*INDOOR GARDENING*

**The Hobby Greenhouse Association** has been promoting indoor gardening since 1976. They have a members-only magazine and newsletter:

hobbygreenhouse.org

*INSECT IDENTIFICATION*

**Iowa State University** sponsors a bug-identification site:

bugguide.net

**North Carolina State Extension** has a Pest Management link:

growingsmallfarms.org/Photos of Pest Insects

**Google** has information and pictures of North American insects, spiders, and bugs; just type the name in the site's BugFinder:

insectidentification.org

*INTERNET PLANT SITES*

**Friends of the Garden International** has a list of plant sites dedicated to plant identification and information:

friendsofthegarden.org>Plant Sites

*INVASIVE PLANTS*

**The Weed Science Society of America** has a huge database of invasive plants:

wssa.net/Weeds/Invasive Plants

**The USDA National Agricultural Library** has a complete profile of invasive plants, plus databases and PDF files:

invasivespeciesinfo.gov/plants/main.shtml

*LAWN CARE*

**The Lawn Institute** features lawn care and the recommended varieties of grass for your area:

thelawninstitute.org.

*MULCH AND MULCHING*

**The California Department of Resources Recycling and Recovery** has a PDF download "A Landscaper's Guide to Mulch: Save Money, Control Weeds, and Create Healthy Landscapes":

calrecycle.ca.gov/search

**The Chicago Botanic Garden** has a PDF fact sheet "Mulch" and a question and answer forum concerning mulch:

chicagobotanic.org/plant info/mulch

***NATIVE PLANTS*** *(see also BEES, BUTTERFLIES, AND POLLINATORS; BIRDS; WILDFLOWERS)*

**The National Wildlife Federation** has a Native Plant Finder that shows, based on your zip code, the best native plants to attract butterflies and moths, and support birds and other fauna:

nwf.org/Native Plant Finder

***ORGANIC GARDENING BASICS***

**Rodale, Inc.** has been a pioneer in the organic gardening movement for 60 years:

RodalesOrganicLife.com/Garden/Beginners-Guide-Organic-Gardening

***PEST CONTROL AND PESTICIDES***

**The US Environmental Protection Agency** has a PDF download "Citizen's Guide to Pest Control and Pesticide Safety Guide," publication ID 262:

publications.usa.gov/262

***PESTICIDE INFORMATION***

**The EXtension TOXicology NETwork** has objective, science-based Pesticide Information Profiles (PIPs) on a wide variety of pesticides:

extoxnet.orst.edu

***PLANT DISEASES***

**Cornell University** has a Plant Disease Diagnostic Clinic with a complete list of plant disease fact sheets and photos, plus links to other sites:

plantclinic.cornell.edu

***PLANT HARDINESS ZONE MAP***

**The US National Arboretum** has an interactive GIS map that scales down to one-half mile, plus a zip code finder for all US zip codes:

usna.usda.gov/Hardzone

*PLANT IDENTIFICATION*

**The USDA Natural Resources Conservation Service** has a comprehensive plant database of over fifty thousand images:

plants.usda.gov/Checklist

**Friends of the Garden International** has a list of all the plant sites dedicated to plant identification and information:

friendsofthegarden.org>Plant Sites

**The free plant identifier apps for mobile devices/smartphones** have built in image recognition technology:

Garden Answers; Google Goggles; GreenSnap; Leafsnap; Like That Garden; iPlanzen; Picture This; Plantifier; PlantChecker; What's that flower?

*PLANTING DATES*

**The *Old Farmer's Almanac*** is a yearly periodical that features planting charts:

Almanac.com/Gardening/Planting-Dates

*PLANT SOCIETIES, CLUBS, AND ORGANIZATIONS*

**The American Horticultural Society** has a list of all the national plant societies in the United States:

ahsgardening.org>Gardening>Resources

*POISONOUS PLANTS*

**Cornell University's Dept. of Animal Science** has a database of plants poisonous to livestock and other animals:

poisonousplants.ansci.cornell.edu

**The ASPCA** has a Toxic and Non-Toxic Plants List for dogs and cats:

aspca.org/pet-care/animal-poison-control/toxic-and-non-toxic-plants

**The Canadian government** has a Poisonous Plants Information System on plants poisonous to livestock, pets, and humans:

cbif.gc.ca/Poisonous Plants

**The Weed Science Society of American** has a complete list of plants poisonous to humans:

wssa.net/Weeds/Poisonous Plants

*RASPBERRIES AND BLACKBERRIES*

**The North American Raspberry and Blackberry Association** is for gardeners, wannabe producers, and commercial growers:

raspberryblackberry.com

*ROCK GARDENS*

**The North American Rock Garden Society** is for gardening enthusiasts interested in growing true alpine and rock garden plants:

nargs.org.

*ROSES*

**The American Rose Society** is the oldest single plant horticultural society in American. It offers guides, information, and interesting links, plus links to other gardening societies.

roses.org

*SEEDS*

**Native Seeds of the Southwest** is a nonprofit seed conservation organization for seeds from the southwestern US and northwestern Mexico. It features open-pollinated, GMO-free seeds of heirloom and landrace varieties:

nativeseeds.org

**Seed Savers Exchange** is a nonprofit seed bank that maintains a variety of heirloom, organic, and open-pollinated garden seeds. They have a free one-hundred-page seed catalog:

seedsavers.org

**Seeds of Diversity** is a Canadian nonprofit seed bank and seed library that provides detailed, step-by-step seed saving and storage information. The site contains a Canadian Seed Catalog Index:

seeds.ca

**Southern Exposure Seed Exchange** is a worker-run cooperative that offers more than seven hundred varieties of organic and open-pollinated vegetable, flower, herb, grain, and cover crop seeds. They have a free catalog:

southernexposure.com

*SUNDIALS*

**The North American Sundial Society** shows how to make and use this oldest clock in the world:

sundials.org

*VEGETABLE DISEASES*

**Cornell University's Plant Pathology** web page has vegetable disease fact sheets that list diseases by crops, along with color photos and a diagnosis:

vegetablemdonline.ppath.cornell.edu

*VEGETABLE GARDENING*

**The National Gardening Association** has four informative articles: "Vegetable Gardening 101," "Growing Vegetables," "Gardening Guide for Beginners," and "Planning a Vegetable Garden":

garden.org

*WATER CONSERVATION*

**The University of California Cooperative Extension** has a PDF download "Water Conservation Tips for the Home Lawn and Garden":

ucanr.edu/PDF 47994

**The US Environmental Protection Agency** has two PDF downloads "Water Efficient Landscaping" and "Green Scaping: The Easiest Way to a Healthy Yard":

publications.usa.gov>Going Green>Water Conservation

*WEEDS*

**The Weed Science Society of America** has a complete database of garden weeds and their images, plus resources for managing weeds in the lawn and garden:

wssa.net/Weeds/Identification/Garden Weeds

**The University of Wisconsin's weed identification site** has an all-inclusive Weed Identification Tool:

weed.idwisc.edu

**The University of Missouri's Division of Plant Sciences** produces a free download weed app for mobile devices/smartphones:

idweeds

***WILDFLOWERS***

**The Ladybird Johnson Wildflower Center** located at the University of Texas has a native plant database of over eight thousand plants. The site's free "Mr. Smarty Plants" has answered more than ten thousand plant and gardening questions to date:

wildflower.org/plants

**Friends of the Wild Flower Garden** is the oldest wildflower garden in the United States. The site has links, information, and a photo gallery of plants of all seasons:

friendsofthewildflowergarden.org

Dear Reader,

Thank you for picking up this book. If you found it helpful in any way, could you please consider posting a short review? It would be gratefully appreciated.

## BIBLIOGRAPHY


Allen, Laura. *Greywater, Green Landscape: How to Install Simple Water-Saving Irrigation Systems in Your Yard.* North Adams, MA: Storey Publishing, 2017.

Bainbridge, David A. *Gardening with Less Water: Low-Tech, Low-Cost Techniques; Use up to 90% Less Water in Your Garden.* North Adams, MA: Storey Publishing, 2015.

Bills, Jan Coppola. *Late Bloomer: How to Garden with Comfort, Ease, and Simplicity in the Second Half of Life.* Pittsburgh: St. Lynn's Press, 2016.

Bucks, Christine, and Fern Marshall Bradley, eds. *The Resourceful Gardener's Guide.* Emmaus, PA: Rodale Publishing, 2001.

Clausen, Ruth Rodgers, and Thomas Christopher. *Essential Perennials: The Complete Reference to 2700 Perennials for the Home Garden.* Portland, OR: Timber Press, 2015.

Cutler, Karan Davis. *Burpee: The Complete Vegetable and Herb Gardener: A Guide to Growing Your Garden Organically.* Hoboken, NJ: John Wiley & Sons, 1997.

Easton, Valerie. *The New Low-Maintenance Garden: How to Have a Beautiful Productive Garden and the Time to Enjoy It.* Portland, OR: Timber Press, 2009.

Gillman, Jeff, and Meleah Maynard. *Decoding Gardening Advice: The Science Behind the 100 Most Common Recommendations.* Portland, OR: Timber Press, 2012.

Gilmer, Maureen. *The Budget Gardener: Twice the Garden for Half the Price.* New York: Penguin, 1996.

Green, Charlotte. *Gardening Without Water: Creating Beautiful Gardens Using Only Rainwater.* Tunbridge Wells, UK: Search Press, 1999.

Greenwood, Pippa. *Pippa Greenwood's Gardening Year: The Busy Gardener's Essential Month-by-Month Guide.* London: Headline Book Publishing, 2004.

Hupping, Carol, Cheryl Winters Tetreau, and Roger B. Yepsen Jr. *Rodale's Book of Hints, Tips & Everyday Wisdom.* Emmaus, PA: Rodale Press, 1994.

Lancaster, Roy. *Perfect Plant, Perfect Place: The Surest Way to Select the Right Outdoor and Indoor Plants.* New York: DK Publishing, 2010.

MacLeod, Jean B. *If I'd Only Listened to My Mom, I'd Know How to Do This: Hundreds of Household Remedies.* New York: St. Martin's Griffin, 1997.

Massingham, Rhonda Hart. *The Dirt-Cheap Green Thumb: 400 Thrifty Tips for Saving Money, Time and Resources.* North Adams, MA: Storey Publishing, 2009.

Millard, Elizabeth. *Indoor Kitchen Gardening: Turn Your Home Into a Year-round Vegetable Garden.* Minneapolis: Cool Springs Press, 2014.

Philip Lief Group. *National Gardening Association Dictionary of Horticulture.* New York: Penguin Books, 1994.

Oster, Doug, and Jessica Walliser. *Grow Organic: Over 250 Tips and Ideas for Growing Flowers, Veggies, Lawns and More.* Pittsburgh: St. Lynn's Press, 2007.

Pollan, Michael. *In Defense of Food: An Eater's Manifesto.* New York: Penguin Press, 2007.

Robinson, Jo. *Eating on the Wild Side: The Missing Link to Optimum Health.* Lake Dallas, TX: Helm Publishing, 2013.

Stout, Ruth. *Gardening Without Work: For the Aging, the Busy, and the Indolent.* New York: Devin-Adair, 1961.

Printed in Great Britain
by Amazon

34141311R00099